SO-BOE-525

Zoology
49 Science Fair Projects

Other books in the series

Insect Biology: 49 Science Fair Projects
H. Steven Dashefsky

Botany: 49 Science Fair Projects
Robert L. Bonnet and G. Daniel Keen

Botany: 49 More Science Fair Projects
Robert L. Bonnet and G. Daniel Keen

Computers: 49 Science Fair Projects
Robert L. Bonnet and G. Daniel Keen

Earth Science: 49 Science Fair Projects
Robert L. Bonnet and G. Daniel Keen

Environmental Science: 49 Science Fair Projects
Robert L. Bonnet and G. Daniel Keen

Microbiology: 49 Science Fair Projects
H. Steven Dashefsky

Space and Astronomy: 49 Science Fair Projects
Robert L. Bonnet and G. Daniel Keen

Zoology
49 Science Fair Projects

H. Steven Dashefsky

Illustrations by Debra Ellinger

TAB Books
Division of McGraw-Hill, Inc.
New York   San Francisco   Washington, D.C.   Auckland   Bogotá
Caracas   Lisbon   London   Madrid   Mexico City   Milan
Montreal   New Delhi   San Juan   Singapore
Sydney   Tokyo   Toronto

pbk   1 2 3 4 5 6 7 8 9 0 FGR/FGR 9 9 8 7 6 5 4
hc   1 2 3 4 5 6 7 8 9 0 FGR/FGR 9 9 8 7 6 5 4

Product or brand names used in this book may be trade names or trademarks. Where we believe that there may be proprietary claims to such trade names or trademarks, the name has been used with an initial capital or it has been capitalized in the style used by the name claimant. Regardless of the capitalization used, all such names have been used in an editorial manner without any intent to convey endorsement of or other affiliation with the name claimant. Neither the author nor the publisher intends to express any judgment as to the validity or legal status of any such proprietary claims.

Library of Congress Cataloging-in-Publication Data
Dashefsky, H. Steve.
    Zoology : 49 science fair projects / by H. Steven Dashefsky.
        p.   cm.
    Includes index.
    ISBN 0-07-015682-4 (H)   ISBN 0-07-015683-2 (P)
    1. Zoology projects. I. Title.
QL52.6.D37 1994
591'.078—dc20   94-12060
                                                              CIP

Acquisitions editor: Kimberly Tabor
Editorial team: Jim Gallant, Editor
                Susan W. Kagey, Managing Editor
                Joanne Slike, Executive Editor
                Joann Woy, Indexer
Production team: Katherine G. Brown, Director
                Ollie Harmon, Coding
                Rose McFarland, Layout
                Nancy K. Mickley, Proofreading
Designer: Jaclyn J. Boone
Technical reviewer: Paul J. Hummer, Jr., Adjunct Professor,   0156832
    Education Department, Hood College, Frederick, Md.   SFP

Disclaimer

Adult supervision is required when working on these projects. All the projects within this book assume your teacher or some other knowledgeable adult advisor assists you throughout the project. No responsibility is implied or taken for anyone who sustains injuries as a result of using the materials or ideas, or while performing the procedures put forth in this book.

Use proper equipment (gloves, safety glasses, protective clothing), wear clothes and shoes that are appropriate for the science lab, and take other safety precautions. Read and follow the manufacturers' instructions when using chemicals and equipment. Use chemicals, dry ice, boiling water, flames, or any heating elements with extra care. Wash hands after project work is done. Taste nothing. Tie up loose hair and clothing. Follow step-by-step procedures and avoid shortcuts. Never work alone. Additional safety precautions are mentioned throughout the book. If you use common sense and make safety a first consideration, you will create safe, fun, educational, and rewarding projects.

Contents

Acknowledgments

The author wants to thank Mr. Eugene Ruthiney and Dr. Nathan Dubowsky at Westchester Community College in Valhalla, New York, and Abalardo Moncayo, a Ph.D. candidate at the University of Massachusetts, Department of Entomology, for many of the projects in this book.

Safety & supervision

All the projects in this book require adult supervision. Don't assume the organisms, equipment, and chemicals you are working with are harmless. Follow all instructions.

The entire project should be read and reviewed by the student and the supervising adult before beginning. The adult should determine which portions of the experiment the student can perform without supervision and which will require supervision.

Symbols used in this book

The following icons are used throughout the book to help adults identify which experiments might require closer supervision for younger students.

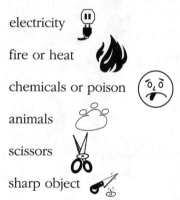

electricity

fire or heat

chemicals or poison

animals

scissors

sharp object

How to
use this book

The best way to select a project is to find out what interests you about animals, if you don't already know. Did you ever wonder what hidden animals could be discovered in a handful of dirt or a cupful of pond water? Are you interested in learning more about your pets' lives? Can you teach your hamster something or see blood circulation in your fish's tail?

Stop and let your mind wander for a while. What comes to mind? It could be anything, anywhere, anybody! Once you've opened your mind and let your imagination run wild, look through the table of contents in this book for more specific topics to research. This book contains 49 science fair projects about animals.

For more information, read the Introduction to each of the nine sections in this book. Then, read the Overview of each project for additional information. Select a project that you are not only interested in, but about which you are truly enthused and wish to learn more.

Each of the 49 projects has an Overview, Materials, Procedures, Conclusion, and Going Further section. Each experiment gives you step-by-step instructions, but leaves you to draw your own conclusions from the data you collect.

Overview

The Overview section gives background information about the topic, explains the purpose of the experiment, and poses questions about the topic. These questions will help you develop a hypothesis for your experiment. (Developing a hypothesis and the scientific method are discussed in more detail later in this introduction.)

Materials

The Materials section lists everything needed to perform the experiment, but you can improvise when necessary. The following materials are not listed, but should be available for any of the projects: a pad and pencil for note taking, Scotch tape, scissors, and water.

Procedures

The Procedure section gives the step-by-step instructions on how to perform the experiment and suggests how to collect data. Be sure to read through this entire section before undertaking any project.

Conclusion

The Conclusion section doesn't draw any conclusions for you. Instead, it asks questions to help you interpret the data and come to your own conclusions.

Going Further

The Going Further section is an important part of every project. It lists many ways for you to continue researching the topic beyond the original experiment. Suggestions are given on what to read and what additional experimentation can be performed. Performing some of these suggestions can assure that the topic has been thoroughly covered and show you how to broaden the scope of the project. The best way to assure an interesting and fully developed project is to include one or more of the suggestions from the Going Further section. Combining related projects is an excellent way to adapt these projects for older students.

SCIENCE FAIR GUIDELINES

Most science fairs have formal guidelines or rules. For example, there may be a limit to the amount of money spent on a project or the use of live organisms. There might be regulations on the use of certain animals. Many fairs limit the use of vertebrates (animals with a backbone). Be sure to review these guidelines and check that the experiment poses no problems.

BE SURE TO USE THE SCIENTIFIC METHOD

Science fairs give you the opportunity to not only learn about a topic, but to participate in the discovery process. Although you probably won't discover something previously unknown to mankind (although you never can tell), you will perform the same process with which discoveries are made. The scientific method is the basis for all experimentation. It simply, yet clearly, defines what scientific research is all about. The scientific method can be divided into five steps.

Purpose

What question do you want to answer, or what problem would you like to solve? For example, "Can shaking hands transfer pathogens from one person to another, and if so, how many people can be contaminated?" The Overview section of each project gives questions and problems to think about.

Hypothesis

The hypothesis is an educated guess, based on preliminary research, that answers the question posed in the purpose. You might hypothesize that pathogens can be transferred by a handshake and that three people can be contaminated. (See Project 18.)

Experimentation

The experiment determines whether the hypothesis was correct or not. If the hypothesis wasn't correct, a well-designed experiment would help determine why it wasn't correct.

There are two major parts to the experiment. The first is designing and setting up the experiment. How must the experiment be set up and what procedures must be followed to test the hypothesis? What materials will be needed? What cultures, if any, are needed? What step-by-step procedures must be followed during the experiment? What observations and data must be made and collected while the experiment is running? Once these questions have been answered, the actual experiment can be performed.

The second part is performing the experiment, making observations, and collecting data. The results must be documented (written down) for study and analysis. The more details you record, the better.

Research

This part of the project should begin before starting the project and continue after the results are collected. Read as much as you can about the topic you are studying. Use any and all sources available to you and try to be the expert on the subject. Once the experiment is completed, analyze the results, and see how what you have learned compares with what is already known about the subject.

Conclusion

Once you have collected and analyzed the data, and researched the subject, you can draw your conclusions. Creating tables, charts or graphs will help you analyze the data and draw conclusions from it.

The conclusions should be based upon your original hypothesis. Was it correct or incorrect? If it was incorrect, what did you learn from the experiment? What new hypothesis can you create and test? Something is always learned while performing an experiment, even if it's how "not" to perform the experiment the next time.

PUT YOUR SIGNATURE ON THE PROJECT

All the projects in this book can be used for a science fair project. What will make these projects award winners are the individual ideas that you bring to the project. Read the Going Further section for ideas on how to modify and extend projects. Speak with your teacher, parents, and friends about a project in this book that interests you. What suggestions do they have about modifying or extending a project to make it truly your own?

THE PROJECTS IN THIS BOOK

The projects in this book have been grouped into nine sections: 1) Behavior, 2) Systems, 3) Animals in their Environment, 4) Beyond the Naked Eye, 5) Animal Lives, 6) Communications and Senses, 7) Growth and Development, 8) Form and Function, and 9) Applied Science.

There is an introduction at the beginning of each section that briefly describes the projects in that section. The Overview in each project details more about that particular project.

Introduction to zoology

This book contains experiments about animals of all sizes and shapes. Zoology, the science of animal life, is concerned about animals from 1/8,000th of an inch long, such as the parasites that cause malaria, to those over 100 feet in length, such as the whale.

Approximately 1.5 million types of organisms (plants and animals) have been identified on our planet and more than 1 million of them are animals. Animals are found almost everywhere on our planet. They can be found on and in the land— from the elephant stomping across the savanna to a community of millions of microbes living in the topsoil.

They are found in almost all bodies of water—from zooplankton in ponds, to mosquitoes that develop in puddles, to large predatory fish and mammals living in the open oceans. Some fly through the air, such as birds, while others float to their destinations on the wind currents.

Animals feed on plants and on each other. Many are parasites that feed on us. Some find their meals in the dead, decaying bodies of what were once plants and animals.

Although the diversity of animals is amazing, the vast majority of all animals are insects, including almost 900,000 species. These arthropods have a skeleton on the outside of their bodies, instead of the inside as we do.

WHAT MAKES AN ANIMAL AN ANIMAL?

Animals are called *consumers*, since they must eat (consume) their food, while green plants, called *producers*, can produce their food. When an animal consumes its food, it gains nourishment that contains stored, chemical energy. The animal digests this food and uses the energy to move, grow, maintain its body, and reproduce. On the other hand, plants can convert sunlight (radiant energy) directly into chemical energy during the process of photosynthesis. This is the primary difference between plants and animals.

Other differences are more obvious. Animals can generally move about freely, while plants cannot. Most animals have some form of a nervous system that allows them to react rapidly to a stimulus, such as an attack from a predator. Plants don't have a nervous system and usually respond to stimuli very slowly.

Finally, animals have a thin, delicate membrane (wall) around all the cells in their bodies, while most plants have a thick, rigid wall that helps support the organism.

THE BASICS OF ANIMAL LIFE

Even though there is enormous diversity in animal life, all must handle the basic functions of living. These basic functions include: 1) ingesting and digesting food so it can be transported to all the cells in the body, 2) respiration to release the chemical energy stored in the food, 3) responding to stimuli, such as light, heat, or an attacker, 4) removal of the waste products it produces while performing functions (1) and (2) above, and finally, 5) reproduction, to continue the species.

(1) Ingesting, digesting, and transporting food can be as simple as bringing a food particle through the cell membrane, adding a few chemicals, and dispersing it throughout the cell, as in an amoeba. In higher forms of life, such as mammals, it means a complex digestive system to get the food into the body and into a usable form, and a circulatory system to distribute the nutrients to all the cells throughout the body.

(2) Respiration is the biochemical process of breaking down sugar molecules to release the chemical energy stored within, so it can be used by the animal to survive. (Respiration is not unique to animals. Plants perform respiration as well as photosynthesis.) Animals use the released energy to move, transport substances within their bodies, grow and maintain their body tissues, and reproduce. Some even use it to produce light, such as in fireflies.

(3) Animals must respond to their environment. Whether it is a one-celled animal moving away from light or a hyena attracted to a rotting carcass, some form of nervous system controls their bodies, allowing them to respond appropriately.

(4) What an animal eats, but doesn't use, becomes waste. Most biochemical processes produce waste products as well. All this waste cannot be left to accumulate in the animal's body or it would poison itself. Simple one-celled animals only have to pass wastes out through their cell membranes, but more complicated forms of life require specialized organs. These organs, such as our kidneys and bladder, help remove wastes from the cells and from our bodies.

(5) Finally, animals must reproduce if the species is to survive. Some simple, one-celled animals reproduce by fission, meaning they simply divide into two individuals. Others can create new individuals by budding. Still others produce young asexually without the need of two sexes. Most higher forms of animals, however, reproduce sexually by the joining of cells produced by individuals of opposite sexes.

THE STUFF OF LIFE

Animals are made up mostly of water. The amount of water ranges from about 50% to 90%. This is because our cells are filled primarily with water and animals consist primarily of cells. Cells are filled with a fluid called *cytoplasm*, which is composed of proteins, carbohydrates, fats, salts, and, of course, a lot of water.

All the materials found in living cells can be found in the nonliving world. These chemicals pass from the rocks, soil, water, and air, into living organisms, primarily by the green plants during photosynthesis. These substances are then passed through food chains and food webs to all forms of life, only to be returned to the nonliving world when the organisms die and decompose.

THE ANIMAL KINGDOM AND THE CLASSIFICATION SYSTEM

Scientists classify all forms of life according to how closely related they are to one another. You can compare this classification system to a file cabinet containing many drawers, with each drawer containing many folders and dividers. Each organism is placed on its own sheet of paper that must be placed somewhere in this file cabinet.

The more closely related two types of animals are, the more closely their sheets of paper will be placed in the cabinet. For example, an insect and a dog would be placed in two different drawers, since they are so different. However, a dog and a cat would appear not only in the same drawer, but within the same folder, since they are so similar.

To continue this comparison, imagine there are two file cabinets: one is labeled "Plants" and the other "Animals." Let's concentrate on the animal cabinet. Imagine there are 10 drawers in this cabinet. Each drawer represents a "phylum" (phyla is plural). Each phylum contains animals that have certain characteristics in common. Insects and spiders are found in one of the drawers, labeled "Arthropods," since they both have a hard skeleton-like shell called an *exoskeleton*. Dogs and cats are found in another drawer, labeled "Chordata," since they have a backbone.

Inside each drawer are a series of folders with labels. These folders represent "classes" of organisms that have even more similar characteristics. For example, our insect and spider both go into the same drawer (phylum), but will go into different folders within the drawer. The insect is placed in a folder labeled "Insecta," which contains all the animals with six legs (Fig. I-1 top). The spider, however, is placed a folder labeled "Arachnida," since it has eight legs (Fig. I-1 bottom).

Our dog and cat were both placed in a drawer (phylum) labeled "Chordata." They not only go in the same drawer, but are also placed in the same folder (Class), labeled "Mammalia," since they both have fur and mammary glands. Since the dog (Fig. I-2) and the cat (Fig. I-2) are in the same folder, they must be more closely related to each other than the insect and the spider, which are in the same drawer, but different folders.

The classification system continues to distinguish which animals are most closely related by placing them in "orders," "families," "genera," and finally, "species." Species represents only those animals so closely related that they can reproduce among themselves. In our example, the species is represented as a single sheet of paper containing one type of animal.

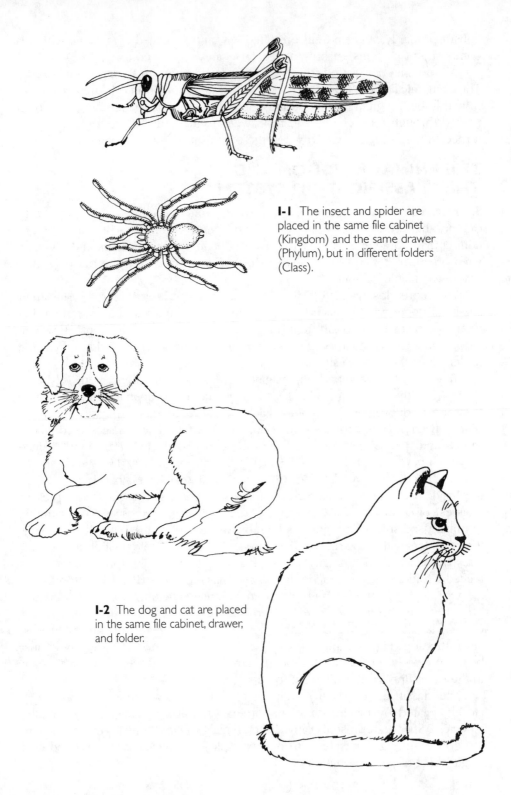

I-1 The insect and spider are placed in the same file cabinet (Kingdom) and the same drawer (Phylum), but in different folders (Class).

I-2 The dog and cat are placed in the same file cabinet, drawer, and folder.

Some of the folders must be thicker than others, since some contain so many more species. For example, the folder labeled "Insecta" must be enormous, since it holds about 900,000 species, while the folder labeled "Arachnida" only holds about 40,000 species.

Do a little reading to see where humans belong in our imaginary filing system. Which file cabinet, drawer, folder, and dividers would contain the sheet of paper containing our species name? Do the same for the first animal you see when you look outside.

Part one

Behavior

The first project investigates pheromones produced by Japanese beetles. If you have ever seen large numbers of Japanese beetles on a rose bush or other plant, you'll know how they all got there after doing this project. The second project helps you determine what kinds of animals living around your home are nocturnal (night) and which are diurnal (day) animals.

The third project helps you to study chemical attractants. What part of a plant emits a chemical that attracts insects to it? The fourth project looks into learning behavior. Can a dog, rabbit, and gerbil all learn how to get past an obstacle to find food?

The fifth project is not about learned behavior, but instinctive behavior. How will fruit flies respond to changes in the direction of gravity? The final project in this section is about the ability to move. Can you distinguish a millipede from a centipede based on its speed (or lack of it)?

I

I can smell for miles
Insects & pheromones

OVERVIEW

Animals use chemical signals to communicate with other members of the same species. These signals are used to mark their territories, attract members of the opposite sex, and to warn members of the same species of danger. Insects are excellent examples of animals that use these chemical signals, called *pheromones*.

The army ant, for example, follows a trail of pheromones that is laid down by the ant ahead of it. This is called a *trail pheromone*. The female gypsy moth releases pheromones to attract males, called a *sex pheromone*. The Japanese beetle uses a pheromone to attract other Japanese beetles to the same area. This is called an *aggregation pheromone*. Most of these pheromones are released into the air and can be detected over very large distances.

Can Japanese beetles attract other Japanese beetles even if they cannot be seen? Will they be able to find other members of their species?

MATERIALS

- Cardboard box (roughly 18" square)
- Two Styrofoam cups with lids
- About 20 Japanese beetles (they can be collected during the warm summer months in the eastern portion of the country)
- Piece of clear plastic or glass, large enough to cover the top of the box
- Scissors

PROCEDURES

Use the scissors to poke many holes in both Styrofoam cups. Place about 10 beetles into one of these cups, cover it with the lid, and label it "beetles." Leave the other cup empty and label it "control" (Fig. 1-1).

Cut off the top of the cardboard box. Then place the cup containing the beetles and the control cup into the box. Cover the large box with the plastic or glass sheet so you can see the cups.

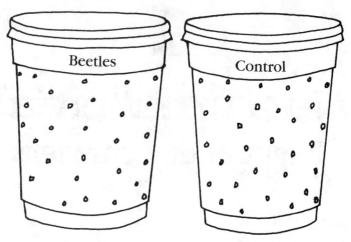

1-1 Poke many holes in both cups so pheromones can be released.

Next, lift up the plastic cover and release the remaining beetles into the large box. Immediately, replace the cover so they do not fly out. Give the beetles some time to get adjusted to their new environment and notice behavior. Record your observations.

CONCLUSIONS

Were the beetles in the box obviously attracted to one of the two cups? Which one? What attracted them to the cup, since they could not see what was inside?

GOING FURTHER

To continue this experiment, make a door on the side of the box. This door should be large enough to allow you to remove the cups. You should be able to fold the door open and shut. Remove the cups through this door and close it so the beetles (in the box) cannot escape. Remove the beetles from the cup and put both cups back into the box. Place the cups up against the walls that were not used in the first part of the experiment. Are the beetles in the box still attracted to the cup that had beetles inside? What can you conclude from this?

Read more about pheromones in insects and other animals.

2
Nocturnal pitfall
Night & day crawlers

OVERVIEW

Organisms that are active during the night are called *nocturnal*, while those active during the day are *diurnal*. Some animals only come out to find food under the cover of darkness, while others are active when the warmth of the sun is available.

If you were to observe the animals that live around your home all day, and then observe them all night, would you see the same or different types of animals? In this project, you'll try to determine if most of the small, crawling animals, such as spiders, insects, crustaceans, and others are diurnal, nocturnal, or active both day and night (Fig. 2-1).

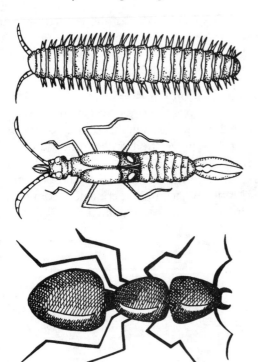

2-1 These are common crawling arthropods that might find their way into a pit trap.

MATERIALS

- Area with at least a small plot of vegetation (a few feet square) to place the trap
- Two similar quart-sized jars, bottles, or tin coffee containers with lids
- Gardener's shovel
- Piece of wood or slate to act as a cover over the trap
- Invertebrate identification guide to identify insects, spiders, and land crustaceans

PROCEDURES

You will create a pit trap that captures small crawling organisms (Fig. 2-2). Locate an area in a garden, lawn, woods, or anywhere around your home where the soil is soft enough to dig a small hole. It is best if the area is surrounded by some vegetation, whether it be grass, shrubs, flowers, etc. Dig a hole just big enough for the entire jar to fit into.

Once the hole is dug, place the jar in the hole so the rim of the jar is just beneath the surface of the soil. Use four similar sized rocks or pieces of brick to support a piece of wood or slate above the submerged jar as you see in Fig. 2-2. This will provide the cover needed for small organisms while they crawl around. It will also keep any rain water from collecting in the jar.

Begin the experiment first thing in the morning. Record the time and the location of the trap and leave the site for 12 hours. After that time, return to retrieve the jar.

2-2 This pit trap can help distinguish nocturnal from diurnal arthropods.

Warning: *Be careful when observing the contents of the jar. There is the possibility that some of the animals caught could be dangerous, such as scorpions and tarantulas.*

Place the lid on the jar and label it "day collection." (If you trapped any animals other than invertebrates, like salamanders or small frogs, record the information and release the animals.) Replace this jar with another empty jar for the nocturnal collection. You will leave this jar in place for 12 hours through the night.

When you return home or to your lab with the day collection, place the jar in the refrigerator for five minutes to slow down the animals. Try to identify as many of the animals as possible. Note how many, as well as the types of organisms, you caught. Record all the results and then release the catch outside.

Return the next morning to collect the nocturnal collection jar. Label this jar "night collection" and repeat the same procedure. If possible, run the experiment in duplicate or triplicate to compare the results.

CONCLUSIONS

Did the day catch differ from the night catch or were they similar? Is the area you tested inhabited with animals that only venture out during the day or during the night? Which animals are nocturnal, which are diurnal, and which are out at all hours?

GOING FURTHER

Research the natural history of the organisms you caught. What do these animals feed on? Are they predators or do they fear becoming prey?

Test a variety of habitats to see if you get varying results.

3
The scent of a rose
Chemical attractants

OVERVIEW

Insects live in a variety of habitats. The main reason an insect chooses a particular place to live is because it provides a source of food. Most insects spend the better part of their lives in search of food.

One of the ways insects track down food is by detecting chemical odors. A mosquito, for example, is attracted to us (one of its food sources) by the carbon dioxide that we breath out.

Many times it is the actual odor of the food that attracts insects. Carrion beetles, for example, are attracted by the ammonia produced in a dead, decaying carcass. Bees are attracted by the smell of flowers.

The Japanese beetle (Fig. 3-1) feeds on flowers, as well. Its primary food source is the rose. What part of a rose plant attracts the beetle? Is it the flower, leaves, stem or some other part?

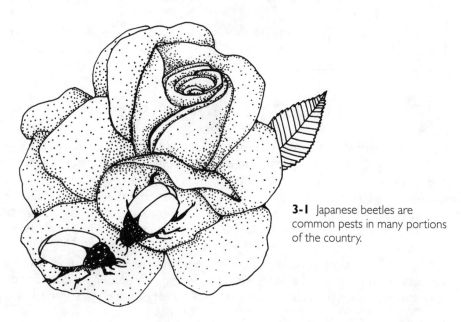

3-1 Japanese beetles are common pests in many portions of the country.

MATERIALS

- Japanese beetles (these can be captured during the warm months in the eastern portion of the country on rose bushes and other types of plants)
- 10 Styrofoam or plastic cups with white lids
- Two live roses including the stem and a few leaves
- Cardboard box about 18" square
- Piece of clear plastic or glass large enough to cover the top of the box so you can see in, but prevents the insects from getting out
- Pencil

PROCEDURES

First, prepare the cardboard box by cutting out of the side a "door" that can be folded open and shut as you see in Fig. 3-2. The door should be large enough for your hand to pass through holding a cup. Remove the cover of the box since it will be covered with the plastic sheet for visibility as you see in Fig. 3-2.

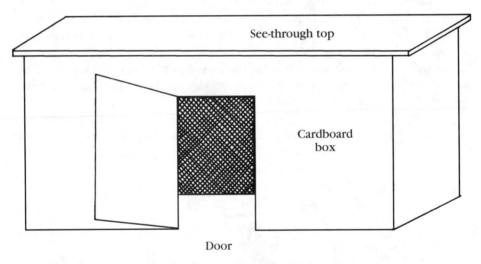

3-2 Cut a door into the side of the cardboard box.

Next, prepare all the cups. Poke many small holes using the pencil throughout the sides of the cups. For the first experimental group, place the flower petals from one of the roses inside one of the cups and cover it with a lid. Label this cup "petals." Place this cup inside the box near the far wall (Fig. 3-3). Place an empty cup with a lid on the opposite wall to act as the control group. Once both cups are ready, place 5 Japanese beetles inside the box and observe their behavior through the cover. Record your observations.

After these observations are complete, remove both cups. Place a few rose plant leaves and part of the stem into a new cup and label it. Take another empty cup with a lid to act as another control. Put these two cups into the box.

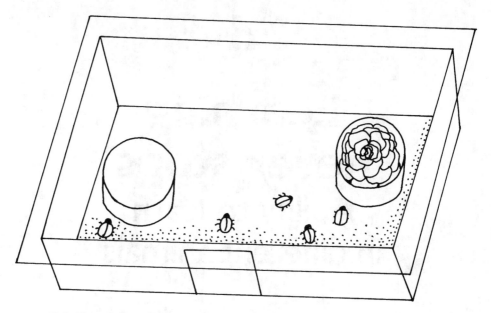

3-3 The cardboard box contains the two paper cups and five Japanese beetles.

(Place them near different walls of the box than the first two cups since the odor of the rose may still be present on the walls.) Once again, observe their behavior and record the results.

For the last part of the experiment, place two new cups, one with rose petals and the other with rose leaves and stems, in the box at the same time. Use a third empty cup as a control. Record your observations.

CONCLUSIONS

Do the Japanese beetles prefer the cup with the rose to the empty cup? Do they prefer the cup with the leaves and stems to the empty cup? Do they prefer the flower or the leaves and stem when given a choice? Can you determine from your observations, what part of the plant actually attracts these beetles?

GOING FURTHER

Investigate what other odors attract Japanese beetles? Continue this experiment by placing different things into the cups that have an odor you think may attract (or repel) the beetles. Always use an empty cup as a control and to see if there is an actual difference.

Learn more about chemical communications among animals.

4
Learning never stops
Ability to learn in different animals

OVERVIEW

The ability to learn new things from past experiences exists in many types of animals. Not only can people learn how to find their way through a maze, but so too can dogs, mice, earthworms, and ants. Of course, the degree to which they learn varies greatly.

In this project, you'll compare the ability of many animals to learn the simple task of figuring out how to obtain food on the other side of a barrier. Can animals such as a dog or cat; rabbit or a ferret; and hamster, gerbil, or mouse learn how to overcome an obstacle to obtain food? How does this ability to learn compare among these animals?

MATERIALS

- Access to at least three pets such as a dog or a cat; a rabbit or a ferret; and a hamster, gerbil, or a mouse
- Wire or wood lattice fence about 6' long and high enough so that the larger animals cannot jump over
- Support to prop up the fence
- Barrier about 8" long and 6" high for the smaller animals plus a support to hold it upright (a small piece of the large barrier mentioned above will work well or a portion of an old window screen can be used)
- Special treat for each animal, such as a dog bone and a yogurt drop for the smaller animals
- Stop watch or watch with a second hand

PROCEDURES

 You will run time trials to see how long it takes for each pet to learn to find the food on the other side of the barrier. The experiments will work best if the animal is hungry. Set up the fence barrier for the larger animal. Hold the dog or cat on one side and show them the treat (Fig. 4-1). Place the treat on the other side in view, release the animal and start the stopwatch.

4-1 A panel of wood lattice acts as a barrier to the food for the larger animals being tested.

Time how long it takes for the animal to find the treat. Then, repeat the same procedure and once again time how long it takes. Repeat this procedure until the animal finds the treat quickly. At this point, you can consider the animal to have "learned" to get the food. Record the number of trials and the times for each.

Perform the next set of trials with a rabbit, ferret, or similar type of pet. Once again show the pet the food and begin the trials. Record the number of trials and the time for each, until the animal finds its way to the food quickly.

Finally, set up the small barrier for the mouse, hamster, or gerbil. Use the same technique as described earlier. Fill in a table similar to Fig. 4-2, with the results of all the animals.

Animal tested	Trial number					Total learning		Notes
	#1	#2	#3	#4	#5	# trials	Total time	
Dog								
Ferret								
Hamster								

4-2 Create a table in your journal similar to this to record your observations.

CONCLUSIONS

How many trials did it take before each animal was able to find the food quickly? What was the total amount of time (in all the trials) necessary for each animal to learn to find the food? Which animals learned the quickest and the slowest? Were all animals capable of learning the task?

GOING FURTHER

Create an experiment to test the learning ability of a worm or ant. Use a simple maze in the shape of a T with some food at one end. Determine if these animals are capable of learning which way to turn for food.

Study the anatomy and physiology of "learning."

5
Which way is up?
Geotaxis & insect behavior

OVERVIEW

Many creatures respond to their environment instinctively, meaning they use "built-in" instructions. They get these built-in instructions from the genes they inherit. All animals have these instinctive behaviors. In addition to instinctive behavior, many organisms can learn from their past experiences. They can then respond to their environment based on what they have learned during their lives. (See the previous project for more on learned behavior.)

An organism's instinctive response to a stimulus is called a *taxis*. There are numerous types of taxes. For example, the response to or from light is called *phototaxis*, while a response to or from touch is called *thigmotaxis*. If the response is away from the stimulus, it is called a negative response, while a positive response means the animal moves toward the stimulus.

The instinctive response of an organism to the earth's gravity is called *geotaxis*. How does the fruit fly (*Drosophila melanogaster*) respond to gravity? Was it born with a built-in negative or positive response to gravity?

MATERIALS

- Vial containing a culture of about 12 fruit flies; for example a 25 cm test tube with a foam stopper to keep the flies from escaping (available from a scientific supply house or they can be caught swarming around a rotting fruit)
- Clamp and stand to hold the vial

PROCEDURES

Attach the clamp to the stand and mount the tube to the clamp in a vertical position as you see in Fig. 5-1. Leave the tube undisturbed for a few moments and then begin your observations. Notice where the flies are in the tube: top, bottom, middle, or dispersed throughout the tube.

Continue these observations for five minutes and then rotate the tube 180 degrees, so the top is now the bottom. As soon as the tube is turned over, once again begin to observe. Watch the movement of the flies. Record your observations by filling in a table similar to what you see in Fig. 5-2. After another five minutes, turn the tube 180 degrees back to the original position and continue your observations.

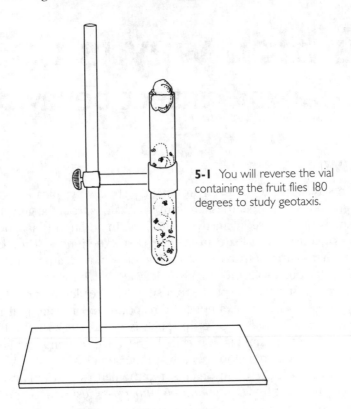

5-1 You will reverse the vial containing the fruit flies 180 degrees to study geotaxis.

Rotation	Vial orientation	Location of flies (time after turn in minutes)										
		1	2	3	4	5	6	7	8	9	10	
—	Cotton plug up	—	—	—	—	—	—	—	—	—	↑	
180°	Cotton plug down	↓↑	↑									
180°	Cotton plug up											

5-2 Create a table in your journal similar to this to record your observations.

CONCLUSIONS

Where were the flies when the vial was in the first position? What happened when you turned the vial over? What behavior did you observe? What happened when you turned it over again, back to the original position?

Do fruit flies move to or away from gravity? Continue this project by researching why they move in the direction they do.

GOING FURTHER

Repeat this experiment for other types of stimuli such as heat or light.

Research other forms of taxis. Why do "bug zappers" attract insects?

6

The tortoise
& the hare
A look at the
millipede & centipede

OVERVIEW

Animals that capture and eat other animals are called *predators*. Animals that eat dead, decaying vegetation are called *scavengers*. The centipede is a predator that captures small insects and other types of arthropods, while the millipede is a scavenger, feeding in and around old rotting logs, leaves, and other vegetation. Although these two types of animals appear at first quite similar, they are really very different. A quick look at the number of legs shows one of the differences.

Do these two types of animals move at the same speed? Can you identify a millipede from a centipede without close inspection, based solely on how fast it moves?

MATERIALS

- Access to areas where you can find millipedes and centipedes (as described in the Procedures section)
- Two jars with lids (for millipedes and centipedes you've collected)
- Area with no vegetation (just soil) at least three feet square
- Forceps
- Piece of string cut to a one foot length
- Stop watch or watch with a second hand

PROCEDURES

First, look at Fig. 6-1 to see the physical difference between a millipede and a centipede. You can find centipedes and millipedes in moist, dark places, such as under rotting logs, or under the bark of a dead tree. (*Be aware that centipedes*

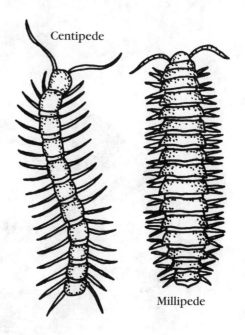

Centipede

Millipede

6-1 With a little practice, it becomes easy to distinguish a centipede from a millipede.

can bite so be sure to use the forceps when collecting or handling them.) Also, look under a layer of leaf litter in the woods, shaded areas of a garden, or even in a basement might be a home to these critters.

Take the foot-long piece of string and the stop watch with you when you go in search of the millipedes and centipedes. Collect at least two or three millipedes and the same number of centipedes. Place them in separate jars. After collecting, find a spot clear of vegetation for the next part of the experiment. (You will be releasing the millipedes and centipedes and will want a clear view where they won't escape.)

Put one of the millipedes in the center of this area and place one end of the string next to it. Have a friend start the stop watch. Extend the string in the direction the animal begins to move (Fig. 6-2). Your goal is to time how long it takes for the animal to travel one foot (to the end of the string). Once the animal has moved the length of the string, gently catch it with the forceps. Record the amount of time it took to travel that distance. Repeat this procedure with the other millipede and then the centipedes.

CONCLUSIONS

Analyze the data and then graph it to illustrate the speed for each type of animal. Then calculate the average for the centipedes and one for the millipedes. Which is faster? Is the difference noticeable? Why do you think one is faster than the other? Can you now identify a millipede or a centipede in the field based solely on how fast it moves?

6-2 Extend the string in the direction the creature is traveling.

GOING FURTHER

Using a dissecting microscope, look at the mouthparts of both animals. Notice the differences between the two and determine why they look they way they do. How do the mouthparts relate to this experiment?

Also, study the number and arrangement of each animal's legs to see how they differ.

Research how millipedes and centipedes defend themselves from predators.

Part two

Systems

Have you ever wondered how scientists put together the skeletal remains of a dinosaur without knowing what it looked like? The first project in this section puts you in this role. The second project investigates when food digestion begins. Does it begin in the stomach? The third project lets you view the circulatory system in action, in the tail of a fish.

The fourth project also deals with the circulatory system, but ours, this time. What is the relationship between heartbeat and pulse? The fifth project leaves the circulatory system and moves on to the nervous system. It concerns the density of nerve receptors in different parts of your skin. Can you sense "touch" on your elbow as well as the palm of your hand? The final project in this section looks into the relationship between exercising and pulse rate.

7
Pellets to skeletons
Reconstructing a skeleton

OVERVIEW

Many museums display the skeletal remains of animals. If the animal is extant (not extinct), it is relatively easy to piece back together, since you can look at a living example of the animal. If the bones are the remains of an animal that existed long ago and is now extinct, there is no living example to guide you. This is like trying to put together a model car or plane without the benefit of seeing a picture of the completed model.

How difficult is it to piece together the skeletal remains of an animal without knowing what the animal looks like? The source of the bones in this experiment will be owl pellets. Owls regurgitate pellets that contain undigested bones of the small animals they have eaten (Fig. 7-1). Investigating the contents of these pellets, locating bones, removing them, and trying to piece them together is similar in many ways to the methods used by paleontologists who discover dinosaur bones and then try to reconstruct the entire skeleton.

Can you piece together the remains found in owl pellets to reconstruct the skeleton?

7-1 Owl pellets contain the indigestible remains of the owl's prey.

MATERIALS

- Sterilized barn owl pellets (from a scientific supply house)
- Fine forceps
- Dissecting probes
- Magnifying glass
- Sheets of black poster board
- Glue
- Reference book that illustrates the skeletal structure of small mammals and birds

PROCEDURES

Many predatory birds capture and eat small animals such as mice, shrews, or small birds. When they eat their prey, they swallow indigestible parts, such as bones and teeth. Instead of passing this waste material out in their feces, they regurgitate pellets that contain this solid waste.

Using forceps, tweezers, and probes from a dissecting kit, carefully tease apart the contents of a pellet. As you find remains, place them on the poster board. Draw sketches as you proceed to help you remember the location of the bones as they were found in the pellet. This will help you piece them back together.

Once you have removed all the remains, try to lay them out on the paper in their proper position (in two dimensional form). (See Fig. 7-2.) When finished, use the reference book to try to identify the animal. Start by looking at the skull, jaw and teeth, if available. Once you've identified the animal, use the illustration of the animal's skeleton to assist you in your reconstruction. Once the reconstruction is as complete as possible, glue the bones to the poster board and label all the parts.

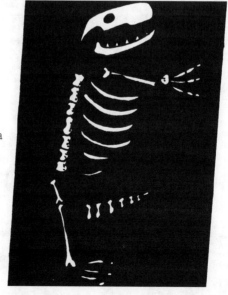

7-2 Lay the remains out on a piece of black poster board.

CONCLUSIONS

What kind of animal remains did you identify? How well were you able to reconstruct the skeleton before identifying the animal in the book? What about after identifying the animal in the textbook? How difficult do you think it is for scientists to put together those dinosaur bones you see in museums?

GOING FURTHER

Read how paleontologists go about their work.

Research the use of dermestid beetles in cleaning the bones of carcasses.

Investigate the digestive processes of predatory animals.

8
Digestion
It begins in the mouth

OVERVIEW

Our digestive system is one long tube running through our bodies. Many processes occur in various parts of this tube: digestion, absorption, and waste removal. Most of the digestion process occurs in the stomach, but when does the process really begin? Does food begin to chemically break down in our mouths before it ever gets to the stomach?

In this project, you will use iodine to test for the presence of starch. If starch is present, iodine will change to a purplish color. The more starch, the more purple. Since a cracker is composed primarily of starch, it should turn iodine purple. When a cracker is digested, the starch breaks down to sugars, which do not cause iodine to change color.

MATERIALS

- A few crackers
- Tincture of iodine (available from a pharmacy)
- Eye dropper or pipette
- Three or four small jars such as baby food jars
- A few disposable tablespoons

PROCEDURES

 First, place a drop of iodine on one of the tablespoons and observe its normal color. Record the color in your notebook. Then, break one cracker into crumbs and place it in one of the jars. Add 4 tablespoons of water and stir until only mush remains. Add 5 drops of iodine to the mush and stir again. (See Fig. 8-1.) Record the color change, if any, that you observe.

Next, thoroughly chew another cracker until it feels like mush (the same consistency as the first jar). Don't swallow! Instead, spit the mush out into a clean jar. This mush consists of the cracker and saliva from your mouth. Add 3 and one half tablespoons of water to this mush and stir. Once again, add 5 drops of iodine and observe the color change, if any.

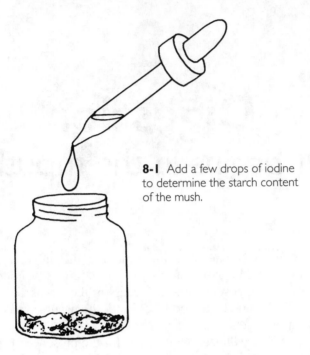

8-1 Add a few drops of iodine to determine the starch content of the mush.

CONCLUSIONS

Does a cracker contain enough starch to change the color of iodine to purple? Does a cracker that has been mixed with your saliva have the same amount of starch? If not, why not? Does the process of digestion begin in your mouth or in your stomach?

GOING FURTHER

Study the biochemical processes that break starch down into sugar (glucose) in the mouth. Continue this study by finding out how this process changes as the food proceeds down your throat, in your stomach, and small intestines.

9
Coming through
Circulation in a fish tail

OVERVIEW

Our circulatory system consists of a heart that pumps blood, arteries that carry blood to the cells throughout our bodies and veins that return the blood to the heart. These arteries become smaller and smaller until they are called *arterioles*. They then become *capillaries*, which are very narrow vessels that deliver the oxygen-rich blood cells to small groups of cells. Capillaries then return the oxygen-depleted cells back through small veins called *venules*. The blood passes through the venules into larger veins until it is finally returned to the heart.

How narrow are the capillaries in the tail of a fish? Do blood cells pass through the capillaries in single file?

MATERIALS LIST

- Large, healthy goldfish (the lighter the color of the tail, the better)
- Cotton balls
- Bottom half of a petri dish or similar dish
- Aquarium water
- Fish net
- Microscope (with high power)

PROCEDURES

Prepare the fish for observation by filling the bottom of the petri dish about one-third with the aquarium water (from the tank containing the fish). Catch the fish with the net and gently lay it in the dish, being careful not to harm it. Immediately cover the entire body, except for the tail, with soaked cotton balls as shown in Fig. 9-1.

Position the dish on the stage of the microscope so the tail is beneath the objective. Begin with low power. Notice the blood vessels. Try to determine which vessels are venules and which are arterioles by the direction of the flow.

Draw sketches of your observations at this magnification and then refocus at the next higher power. Then, move to the highest magnification to observe

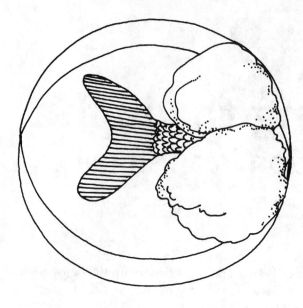

9-1 Viewing blood pass through the tail of a goldfish is one of the simplest, yet most dramatic, ways to gain an understanding of the circulatory system.

actual blood cells passing through the vessels. Draw illustrations of what you see at this magnification. (See Fig. 9-2.)

Don't observe the fish for more than 10 minutes. If you need more time, place the fish back in its tank, wait about 30 minutes, and return to your research. Return the fish to the tank when the experiment is finished.

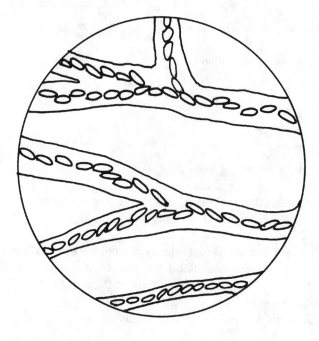

9-2 Can you tell which vessels are venules and which are arterioles by the direction of flow?

CONCLUSIONS

Could you identify venules, arterioles, and capillaries by the direction the blood was flowing? Did you find capillaries with blood cells passing through one-by-one?

GOING FURTHER

Continue the research by studying the actual blood cells. Look at the speed they flow, and their shape and size. What might this tell you about the process of respiration and how it works?

Research the comparative anatomy of circulatory systems. How does this system differ among fish, insects, and a starfish?

Try a similar experiment, but determine the effect of cold or warm temperatures on the circulation of the fish.

10
My beating heart
& pulsing veins
Timing heartbeat & pulse

OVERVIEW

Each contraction of your heart forces blood out of the heart, into the aorta and through the arteries. As the blood is forced through the arteries, the walls of these vessels expand under the pressure, creating a wave that we call a *pulse*. This pulse wave is felt when a nurse or doctor "takes our pulse."

How long does it take for the contraction of the heart (heartbeat) to be felt as a pulse on our wrist? Do they occur simultaneously? A few seconds apart? Ten or 20 seconds apart?

MATERIALS

- Watch with a second hand
- Stethoscope (optional)

PROCEDURES

First, practice listening to the heartbeat of a friend. It would be useful if you can borrow a stethoscope from a doctor for this experiment. Each heartbeat sounds like two thumps. You are interested in the first thump, since this is the contraction of the left ventricle that forces the blood through the arteries.

Once you are familiar with the heartbeat, practice feeling your friend's wrist (radial) pulse as you see in Fig. 10-1. Have your friend hold the stethoscope in place for you. Position yourself so you can hear the person's heartbeat, take the person's pulse, and see the watch at the same time.

While watching the second hand on the watch, listen for the beginning of a heartbeat and feel for the pulse. Can you time the difference between the two? Try this for a few heartbeats and then switch places with your friend.

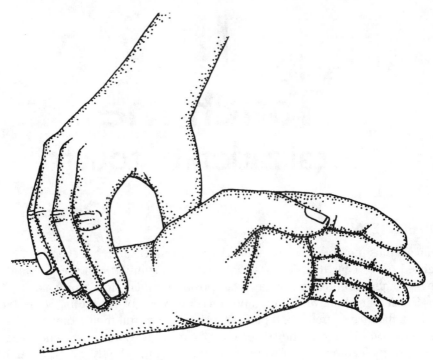

10-1 Feel your friend's pulse by holding his or her wrist as illustrated here.

CONCLUSIONS

Were you able to detect the duration between heartbeat and pulse? Do they occur at almost the same time or is there a lapse between the two? If they appear to occur at almost the exact same time, research how this could happen.

GOING FURTHER

Can you devise a similar experiment using other larger animals such as a dog or a horse? (Have an adult assist you with the horse.) Are the results the same? If so, why?

Research blood pressure and the circulatory system. What are the harmful effects of high and low blood pressure?

11
Touch me
Localization of touch

OVERVIEW

Animals must be able to sense their environment to survive. Receptors are specialized nerve cells that detect specific types of stimuli. Some receptors detect smells, others sounds, still others heat, position, touch, light, and color.

Some animals have unique and amazing types of receptors. Some blood-sucking flies have stretch receptors that detect how full their stomach (gut) has become from the blood they've eaten. When the receptor is stretched enough, the fly stops feeding. If you were to cut the receptor, the fly would continue to feed until it explodes! Some moths have specialized receptors capable of detecting the smell of a mate over five miles away.

We have many kinds of receptors on our bodies. Touch receptors are located all over our skin. Some parts of our bodies have more touch receptors than others. Can you determine which parts of our skin are the most densely packed with touch receptors?

MATERIALS

- Pair of scissors with sharp points
- Ruler with millimeter graduations
- A few friends to help

PROCEDURES

When sharp objects such as the points of scissors contacts the skin, touch receptors are stimulated, and a message is sent to your brain. This results in the sensation of touch. If two points are touched near one another, the touch receptors might have trouble distinguishing between them. The more densely packed the receptors, the more likely they will be able to distinguish two points of contact, even if they are close together. If an area has few receptors, two points close together will feel as one.

Have your friend close his or her eyes and put out the palm of the hands as seen in Fig. 11-1. With the scissors completely closed (one point), gently, you do

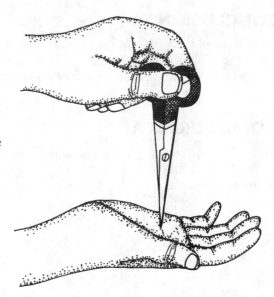

11-1 Begin the test by closing the scissors so there is only one point touching the palm of the hand.

not want to puncture the skin, touch the point to the skin in the middle of the person's palm. Ask the individual how many points are felt and record the data.

Next, open the scissors only a millimeter or two and repeat the test. Progressively, open the scissors in very small increments until your friend can feel the two points. Next, repeat this procedure, but test different parts of the skin. For example, the back of the hand, the fingertip, sole of the foot, elbow, and back of the neck. As you proceed, fill in a table similar to Fig. 11-2.

After completing this procedure, switch places with your friend so you become the test specimen. Repeat the entire procedure mentioned above.

Location	Distinguishing distance
1. Palm	millimeters
2. Back of hand	"
3. Sole of foot	"
4.	"
5.	"
6.	"

11-2 Create a table in your journal similar to this to record your observations.

CONCLUSIONS

Which part of your friend's skin was the most and least sensitive to touch? Do these areas coincide with the tests performed on you? From these results, determine which parts of the body have the most and the least densely packed touch receptors.

GOING FURTHER

Use an anatomy and physiology text to see if your findings are correct.

Study other kinds of receptors found in the skin and try to develop an experiment to test them.

12
Racing pulse
Pulse rate & exercise

OVERVIEW

Each time your heart beats, blood is rapidly forced through the circulatory system to be delivered to all the tissues in the body. The surges of blood through the body can be felt at key points as a pulse. The pulse rate of an individual is an indication of the overall activity of the organism. The more activity, the greater the demand for oxygen, and the faster the pulse rate.

How does exercise affect the pulse rate of humans? How immediate is the effect and how long does it last? How does routine exercise (being "in shape") change these results?

MATERIALS

- Stopwatch or a watch with a second hand
- Someone who exercises routinely and someone who rarely exercises, both about the same age (*Be sure the participants are in sound health and do not have heart problems.*)

PROCEDURES

Have the first participant sit down and not move for two minutes. Take this person's "at rest" pulse rate, using the stopwatch, as you see in Fig. 10-1 (in Project 10). Count the number of beats per minute and create a graph similar to the one you see in Fig. 12-1. Have this person run in place for two minutes and immediately take their pulse again. Continue to take this person's pulse rate every two minutes until it returns to normal (the "at rest" reading). This period is called the *pulse recovery rate.*

Repeat the above steps for all the participants and continue to plot the data on the graph. Next, take readings from other people in varying degrees of "fitness." For example, an athlete and a person who gets no exercise at all, and plot the results in the graph.

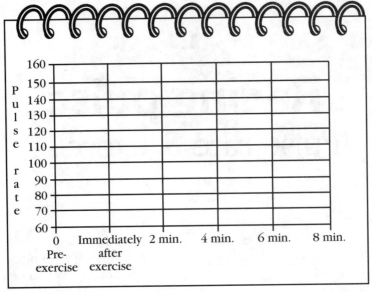

12-1 Plot the data in a graph similar to this one.

CONCLUSIONS

Is there a direct relationship between exercise and the pulse rate for everyone you tested? Is there a difference between those who are more physically fit and those who are not? What's the biggest difference between an athlete and a couch potato? Is it the pulse rate at rest, the pulse rate after running in place, or the amount of time it takes to return to their "at rest" rate (recovery rate) after running in place?

Research why these differences exist? What does this tell you about the advantages of exercising regularly?

GOING FURTHER

Research what other conditions might increase a person's pulse. For example, how does fear (taking a test or hearing a loud noise) affect pulse and why? Relate your findings to this project.

Some athletes show little to no change in pulse rate during routine exercise. Research how this is possible.

Part three

Animals in their environments

All animals must deal with the conditions of their environment. For animals living in northern portions of the country, cold winter with its freezing temperatures is a survival challenge. The first project in this section investigates which stage in a mosquito's development (egg, larvae, pupae, or adult) is most likely to survive the winter?

The second project investigates predator-prey relationships. The praying mantis captures and devours many types of small invertebrates, such as other insects, spiders, and crustaceans, such as the pill bug. Are any of these tasty morsels simply too tough to eat?

We often study an ecosystem by looking at food chains. The third project in this section examines the advantages and disadvantages of using this learning technique. The fourth project investigates the tolerance ranges of aquatic plankton. How can a small change in a single condition, such as pH, affect an entire ecosystem? The final project provides a way to look into the hidden world of earthworms. What do these creatures like to eat? Do they prefer an oak leaf to a kiwi fruit?

13
Baby, it's cold outside
Temperature & mosquito development

OVERVIEW

The development of living things is a very complicated and delicate series of events. So many things can go wrong, it's a wonder that some organisms exist at all. The female mosquito, for example, must find a place to lay her eggs. These eggs are very small and can be eaten by many kinds of predators, or they can be crushed or washed away by water currents.

If the eggs survive and larvae hatch, their main role in life is to eat and grow. The larvae may not survive if there isn't enough food, or they are eaten by predators. If the larvae survive, they turn into pupae. This stage does not feed. Its purpose is to keep from being eaten and to change form (metamorphose) into an adult. Once an adult, the mosquito can fly and leaves the water where it has lived in since birth. The adult must now find a mate, since its primary purpose is reproduction.

All of these stages of development are vulnerable to the environment, but some more than others. How cold can the environment become and still allow mosquito larvae and pupae to survive? Could one of these stages survive temperatures below freezing and still become an adult mosquito?

MATERIALS

- Mosquito larvae all the same size (can be collected in ponds during the summer or ordered from a scientific supply house)
- Three cardboard boxes (roughly 8" square)
- Fine wire mesh that will cover the open top of the box (a window screen will work well)
- Stapler

- Three small plastic cups (no more than 2 inches high and 2 inches in diameter)
- Refrigerator with a freezer
- Aquarium fish food (from a pet store)
- Eyedropper

PROCEDURES

First, prepare the three boxes by cutting off their top flaps. Then, cut out a door on the side that can be folded open and shut, and is large enough for you to put a small plastic cup through. Cover the top of the boxes with the screen mesh and staple it securely in place. (See Fig. 13-1.)

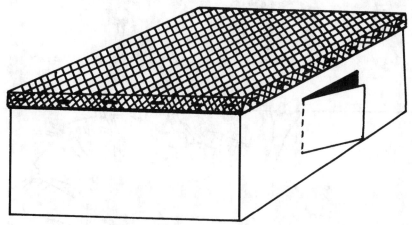

13-1 The cardboard box should be covered with a layer of fine mesh screen, stapled to the top of the sides.

Use an eyedropper to place six mosquito larvae in each of three cups and put a pinch of fish food on the surface of the water. Make sure that all the larvae are approximately the same size. (See Fig. 13-2.) Place one of the cups into each of the boxes. (One cup per box.) Place the first box in the freezer, the second box in the refrigerator, and the keep the third at room temperature in the dark.

The time it will take for the larvae to become adults depends on the age of the larvae at start of this experiment. The larger the larva, the older it is, and the less time it will take to become a pupa, and then a flying adult. Observe the three sets of insects over the next week and see if all three sets develop into adults. Periodically, add the same amount of food to each cup. Record all your observations.

Notice when the larvae change into pupae. (See Fig. 13-3.) When you see adult mosquitoes flying around under the mesh, you can take them outside and release them.

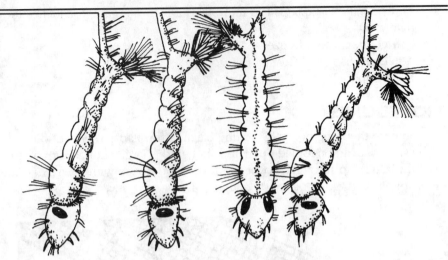

13-2 All of the mosquito larvae must be the same size at the start of the experiment.

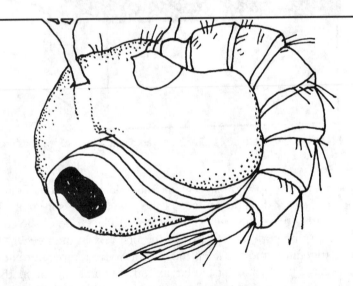

13-3 Mosquito pupae are extremely odd-looking creatures.

CONCLUSIONS

Did all three temperatures allow for the full development of the mosquito? Did they all develop at the same rate? Did any temperature result in the death of the larvae or pupae? Be sure to read the Going Further section, since some of the individuals might appear dead, but not actually be dead.

GOING FURTHER

Continue this experiment to see if the larvae or pupae that did not mature to adults are able to continue their development when removed from the cold and placed in room temperature. At the conclusion of the experiment mentioned above, leave the cups in room temperature for a few days and see what happens. Did any continue to develop and turn into adults? How is this significant in nature?

Read more about the development of mosquitoes.

14
Tough as nails
Predators & prey

OVERVIEW

Many insects are predators, feeding on other animals called prey. There are numerous beetles that devour other insects such as aphids. Many aquatic insect larvae are predators. They grab their prey, insert their piercing-sucking mouthparts into the bodies of the prey, and suck the fluids out of them.

One of the better known insect predators is the praying mantis. It captures prey and devours them with its powerful chewing mouthparts. Praying mantises feed on other insects or relatives called *crustaceans*, such as pill bugs. Both insects and crustaceans are *arthropods*, meaning they have a hard exoskeleton that supports and protects the animal. Some arthropods have harder and thicker exoskeletons than others.

A praying mantis will attempt to capture and devour almost any prey that is small enough not to pose a threat. Most insects such as grasshoppers and smaller crustaceans such as pill bugs are the perfect size to eat. But can a praying mantis always chew through the protective exoskeleton? Do some potential prey have a thick enough coat of armor (exoskeleton) to protect themselves from being devoured?

MATERIALS

- At least one (preferably two) praying mantises (can be caught or purchased from a scientific supply house as egg cases)
- Small live cricket, grasshopper, pill bug, millipede, cockroach, or any other nonpredatory arthropods to feed to the praying mantis (at least four different kinds)
- Terrarium large enough to hold one praying mantis and a few small twigs for it to perch itself upon (you can make one out of a wide mouth jar, using cheesecloth for a top)
- Small plate or bottom of a petri dish
- Cotton balls
- Gloves

PROCEDURES

Place one praying mantis in the jar with a few small twigs and a dish containing a cotton ball soaked with water. (See Fig. 14-1.) Change the cotton every other day and be sure it remains damp throughout the entire project.

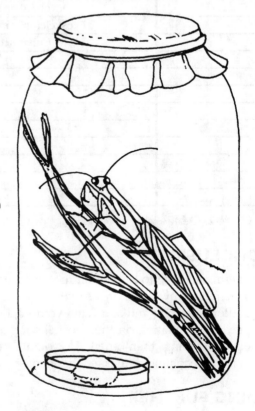

14-1 Be sure to cover the top of the jar so the praying mantis cannot escape, and keep a fresh supply of water available at all times.

Place the first prey into the cage and cover so neither the predator nor the prey can escape. How long it takes before the praying mantis attacks the prey depends upon when it last ate. You don't have to wait to actually see the attack, but return often to see if the mantis is devouring its prey. If you don't see the prey, look around the bottom of the cage for the remains. Record the results.

Place the next type of prey in the cage and make routine observations, once again. Continue this procedure until you've fed the mantis all of the prey and recorded the fate of each encounter. Fill in a table similar to Fig. 14-2. Release the mantis outdoors when you have completed this experiment.

Prey	Final state
1. Cricket	Devoured entirely
2.	Exoskeleton remained
3.	Unharmed
4.	
5.	

14-2 Create a table in your journal similar to this to record your observations.

CONCLUSIONS

Was the mantis able to devour all the different kinds of prey? Did it have trouble penetrating any of the prey's exoskeletons? Was it unable to eat some of the prey? After collecting and analyzing your data, research the anatomy of each of the prey, concentrating on the exoskeleton. Do some arthropods have thicker protective coverings than others? Do some have enough armor to protect themselves from becoming dinner for the mantis?

GOING FURTHER

Dissect a specimen of each type of prey fed to the mantis to actually see differences in their exoskeletons.

Research the chemical structure of the exoskeleton. What makes it so tough?

15

Oh, what a tangled web we weave
Feeding relationships

OVERVIEW

A *food chain* describes the sequence in which organisms serve as food for which other organisms. In other words, it illustrates "who eats whom" in an ecosystem. On land, a typical food chain may appear as follows: grass and trees (*producers*) are eaten by birds, mice, rabbits, and insects (*herbivores*). The herbivores are consumed in turn by frogs, weasels, and foxes (*carnivores*).

When all of the above mentioned organisms die and decay, they release their nutrients back to the soil with the help of *decomposers* (such as worms). The nutrients can once again be taken in by the grass and trees to begin the process over again.

Is there a single food chain that can show all the feeding relationships that exist in any given area? Do food chains really exist in nature or are they purely for educational purposes?

MATERIALS

- Magazines of wildlife including everything from bacteria and algae to plants and large animals
- Scissors
- Paste
- Two large poster boards (as large as you can find, such as 4 feet × 3 feet)
- colored markers

PROCEDURES

Cut out as many animals and plants as you can from the magazines. Then research examples of food chains in many text books and create some of your own examples. Next, you will use all the cut-out animals and plants to create food chains on one of the pieces of poster board.

Before pasting any of the animals to the board, arrange them all into at least 20 different food chains. Each chain should have at least 4 organisms. Once they are all properly arranged, take some blank colored paper and cut out arrows. Place the arrows between the organisms, indicating who is eating who (or what). Mark all members of one food chain with a color code. Once all the cut-outs including the arrows are ready, paste them in place on one of the poster boards as you see in Fig. 15-1. Label this poster, "Food Chains." Make 10–20 food chains, pasting all of them on the poster board.

15-1 The "Food Chain" poster board should display at least 20 examples of food chains side-by-side.

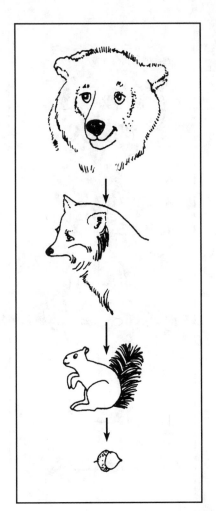 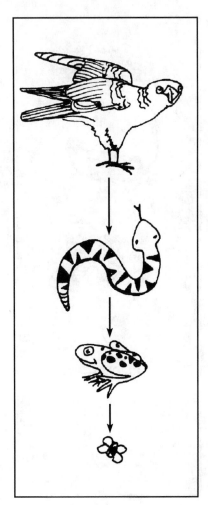

15-1 cont.

Now that the "food chain" board is complete, begin looking for relationships that exist among the food chains. See if any plants or animals from one food chain can serve as food for the adjacent food chain and vice versa. Look for predators and prey, parasites and hosts, plants and animals that feed on those plants. Use a pencil to draw arrows to show these relationships that exist between food chains. Don't forget to look for relationships among food chains on the strips farther away. (See Fig. 15-2.)

Oh, what a tangled web we weave **45**

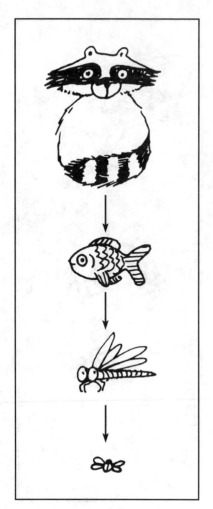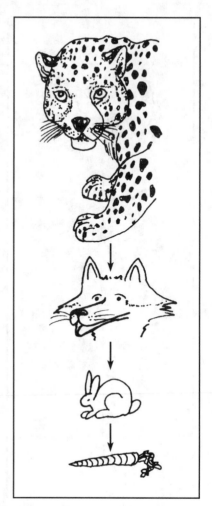

15-1 cont.

After all of the relationships are indicated with arrows, create a "Food Web" chart on the other poster board. Do this by writing a list of all the relationships that you've found. Then, cut up the Food Chain poster board into pieces and re-arrange them on the new chart to indicate all the relationships. You can cut up the food chains into individual pieces if necessary, but don't forget their original relationships.

CONCLUSIONS

What conclusions can you draw from this exercise? Does a food chain actually exist in nature? Are food chains naturally separate from one another? After you have completed this project, does the second poster board look more like a chain or a web? To understand a real world ecosystem, is it easier to study all the relationships that exist or to separate them out into food chains?

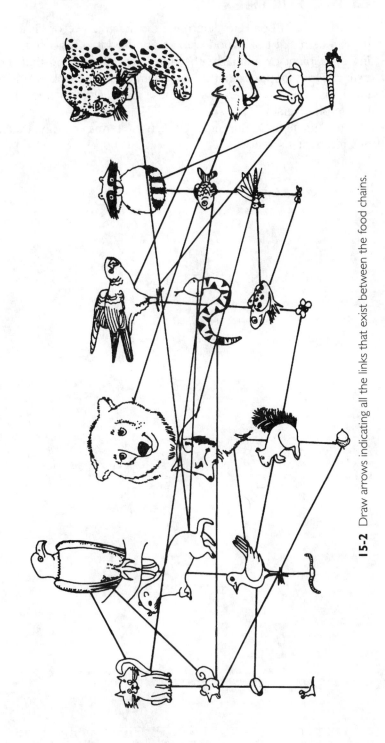

15-2 Draw arrows indicating all the links that exist between the food chains.

Oh, what a tangled web we weave **47**

GOING FURTHER

Use your poster board to determine the advantages (or disadvantages) of an organism being part of a food web as opposed to a food chain. Do this by covering up a plant or animal with white paper indicating that they have become extinct. Does it appear that a food chain or a food web would survive or would it collapse due to the removal of these animals? What does this tell you about a loss in biodiversity?

Investigate the difference between the terms ecology, the environment, and environmental science.

16
Living within the limits
Tolerance ranges of plankton

OVERVIEW

All organisms need certain things to survive. They need certain types and amounts of nutrients, minerals, and gases. They need to live within a certain temperature range and have the right amount of moisture and sunlight available. All of these factors must fall within certain ranges for an organism to survive. These ranges are called *tolerance ranges*. For example, some animals must live in an environment that has a pH (degree of acidity) between 7.3 and 8.4 to survive. This is the pH tolerance range for this animal.

If a factor does not fall within the tolerance range, the organism must move out of the area or it will probably die. If a factor such as pH no longer falls within the tolerance range, it is said to become the *limiting factor*. This means it can limit the success of the species in that habitat.

Plankton are microscopic organisms commonly found in ponds and lakes. (See Fig. 16-1.) Some are free swimming animals (zooplankton) that prey on small microscopic plant life (phytoplankton). Will a change in the water's pH affect the various kinds of plankton in an aquatic environment? How do the populations change when the pH in their environment becomes more basic (less acidic)?

MATERIALS

- Access to a pond or lake
- Two wide-mouth jars or small aquariums
- Two large wide-mouth containers with lids such as rubber tubs (one to collect pond sediment and the other to collect pond water)
- Aquatic microbes (can be collected at the pond or a culture of plankton can be purchased from a scientific supply house)
- Microscope

- pH test paper or meter (with .5 unit accuracy)
- Eyedropper
- Microscope well slides
- Lime or limestone (from a garden center)
- Cheesecloth
- Rubber bands or string
- Plankton identification guide

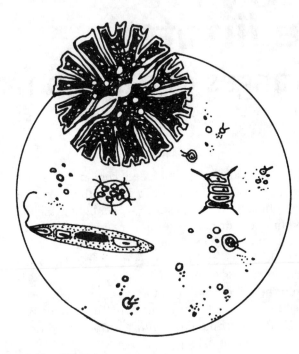

16-1 Examining the pond water under high magnification should reveal a diverse world of plankton and other microbes.

PROCEDURES

Be sure the jars or aquariums are thoroughly cleaned. Rinse them many times so no soapy residue remains. Then, have an adult take you to a pond or lake and use one of the containers to scoop up some mud near the edge. Use the other container to collect enough pond water to fill both aquariums or wide mouth jars as listed in the Materials section. Bring both containers back to the lab or your home.

Place about one inch of the pond mud in the bottom of each aquarium or large wide-mouth jar. Fill each with similar amounts of pond water. Use the pH test paper to measure the pH of the water in both tanks. (They should be almost identical since they came from the same source.)

Label one of the tanks "control" and the other "basic." Take a small sample of water from the "control" jar with an eyedropper, place it in a well slide and observe under the high power objective of the microscope. (See Fig. 16-1.) Use the identification guide to identify the microbes you see. Note the types and numbers of each. (If only a few are present, try collecting at another site or consider purchasing a culture.) Repeat the procedure with the "basic" jar. You should find the same community of organisms since they came from the same source.

Cover the tanks with cheese cloth and hold it in place with string or rubber bands. Place both jars in indirect light, in a warm part of your home or lab for a few days before continuing. After this time, measure the pH level once again. (The pH may have changed, but both tanks should remain similar.)

Next, add a teaspoon of lime to the "basic" tank. (See Fig. 16-2.) Gently stir, wait a few hours and then take another pH reading. If there is no change, add another teaspoon and take another reading. Once you see a visible change in the pH, record the data and wait at least 2 hours. Then, take samples from each tank with an eyedropper and observe under the scope once again. Compare the microbe populations present in the two groups. Continue to raise the pH (with the lime) and observe the microbes in both tanks over a one week period.

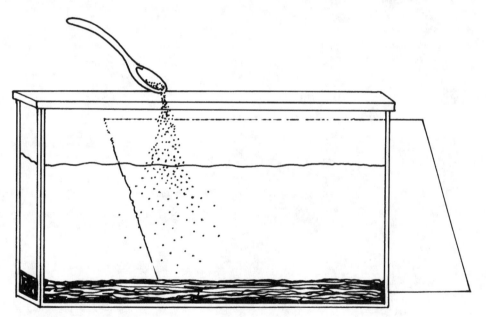

16-2 Pour a teaspoon of lime into the container to raise the pH level.

CONCLUSIONS

The lime should increase the pH (make it more basic). Did a slight increase in the pH level affect the populations of microbes in the aquatic environment? How much of a change in the pH was needed to cause a change in the populations of microbes? If this change occurred naturally, how would it affect the local aquatic ecosystem? Did you change the pH so much that it was no longer within the required tolerance range for some or all of the plankton?

GOING FURTHER

What would happen if rainwater was added to the "basic" tank? Would the ecosystem re-establish itself?

Would you get the same results if you lowered the pH? (**Caution:** *You must work closely with an adult if using acid.*)

Study the latest research on acid rain.

17
Worm dinner
Earthworm eating preferences

OVERVIEW

Millions of earthworms can live in an area roughly the size of a football field. As they burrow their way through the soil, they move, aerate, and enrich the soil. Earthworms ingest the soil, feeding on the organic material found in and on the soil. After munching on their food, the waste passes out of their bodies and is pushed up to the surface. This mixture of waste products and soil that passed through their bodies is called a *worm cast* (Fig. 17-1).

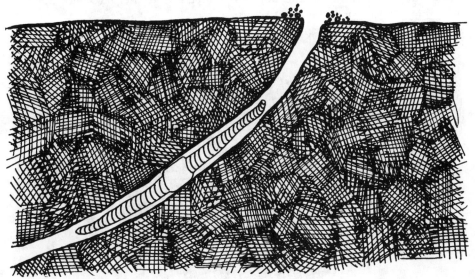

17-1 Earthworms push their waste to the surface. The crumbly material is called worm casts.

Worms will carry food they find on the surface, down through their tunnels to be ingested. What kinds of food do earthworms prefer?

MATERIALS

- Area known to be inhabited by earthworms (look for worm casts in gardens or other areas containing fertile soil)
- Old window screen (you can use old panty hose or similar netting attached to a frame)
- Collection of different kinds of leaves
- Selection of fruits and vegetables (two different kinds of each)
- Large sheet of drawing paper
- Leaf identification guide

PROCEDURES

Prepare a variety of foods to be tested. Do this by collecting at least five different kinds of leaves. Use a guide to identify the leaves. Find some fresh berries from local plants. Also, cut up into a few small pieces (less than ¼" squares) the fruits and vegetables that you've purchased. Lay out all the leaves, berries, and bits of fruits and vegetables in the area known to be inhabited by earthworms.

Place the window screen above the foods and hold it down with either rocks or small spikes. (See Fig. 17-2.) Draw an illustration on the drawing paper of the area beneath the screen so you know exactly where each piece of food is located. Be sure to include on this map the type of food and its location. Include measurements from the edges of the frame to help you locate where everything is (or was) when you return.

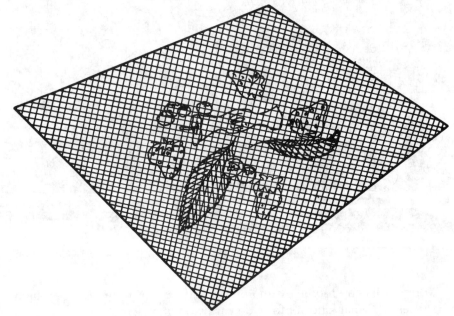

17-2 Cover the various foods with a screen. Draw an illustration so you remember where everything is located.

Return to the plot the next day. Look for earthworm holes and their casts on the ground. Check your drawing to see where everything was originally placed. Look for worm holes and casts. On the map, mark which pieces of food are missing or have been moved. Note the date and time you noticed each piece gone or moved. Don't disturb the plot. Return each day and record the changes. If possible, set up one or two additional plots similar to the first and compare the results.

CONCLUSIONS

Which food disappeared first? Does there appear to be a preferred food for the earthworms? Do they prefer a certain type of leaf? Are there some foods that remained untouched?

GOING FURTHER

Continue this project, but see if the location of the plot changes the results. Prepare the plot near a compost heap, in the woods, or in an open field. Why might the preferred foods be different?

Investigate whether the age of a worm affects its diet.

Part four

Beyond the naked eye

You shake hands with a friend or relative. What's the big deal? You might reconsider the handshake after seeing in the first project how microbes can be passed from one hand to the next, and the next, and the next. In the second project, you'll see what goes on in your body everyday, beyond the naked eye. The food you eat contains starch, but our bodies cannot use the energy locked in these starch molecules. This project has you watch as starch is broken down into useable sugars.

The third project has you investigate our blood cells, antigens and antibodies. If you don't understand what it means to be a certain blood type (O or A for example), you will after performing this project. The fourth project doesn't look at single cells within a large organism, but at a single-celled organism. How does a one-celled animal such as a paramecium ingest and digest its food?

The final project in this section returns to the processes that go on in almost all cells, but at the invisible molecular level. This project has you observe the end results of osmosis.

18

Hazardous
to your health
The biology of an infection

OVERVIEW

Certain illnesses, like the flu, seem to appear at certain times of the year. Although it would appear that the seasons somehow cause these diseases, they are really due to the transmission of disease causing microbes called *pathogens*. Can microbes be easily transmitted from one person to another by a simple handshake? Can people be exposed to these microbes indirectly? If so, how many people can be "contaminated" from a single source?

In this project, you will use harmless yeast cells instead of a pathogen. If you can demonstrate that a handshake transmits a harmless yeast cell from one hand to another, it might be likely that it can also transmit pathogens.

MATERIALS

- Package of bakers or brewers yeast (packaged yeast is available in most supermarkets)
- Sugar
- Pond or tap water
- Large beaker (about 600 ml)
- 15 Sabouraud agar plates (available in prepared form or they can be made from a mix; both are available from scientific supply houses)
- Box of disposable surgical gloves (from a scientific supply house)
- Box of sterile cotton swabs
- Eyedropper
- Incubator (optional)
- At least five individuals (including yourself)
- Marker to mark glass or plastic

PROCEDURES

First, prepare an active culture of yeast cells by adding 50 grams of sugar to 450 ml of tap or pond water in a large beaker. Then, add the contents of one bag of Fleishman's yeast, or a teaspoon of bakers or brewers yeast to the beaker. Stir the liquid and leave the beaker in a warm part of the lab or your home for 24 to 48 hours. Do not cover the beaker. After this time, your beaker should contain an active culture, loaded with yeast cells.

Label three sets of petri dishes #1 through #5, as you see in Fig. 18-1. Have each of the five people put a surgical glove on their right hand. Each person should be assigned a number with you being number 1. Place a few drops of the yeast culture on your glove using the eyedropper. Rub the liquid all around your glove with a sterile swab. This contaminates your glove with this microbe. You are now the source of our simulated pathogen.

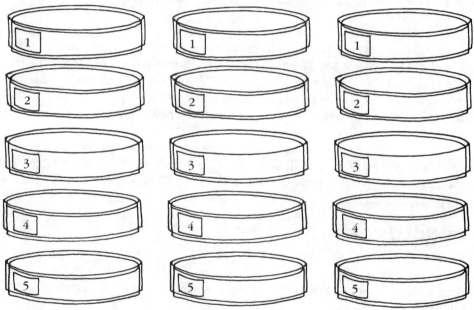

18-1 The experiment will be run in triplicate so you'll need three sets of five petri dishes.

Now, shake hands with the number 2 person. (See Fig. 18-2.) The number 2 person then shakes hands with the number 3 person. This is repeated until the fourth person has shaken hands with the number 5 person.

As soon as all the hand shaking is completed, have each person touch the fingertips of their gloves to one of the properly labeled petri dishes. The number 4 person, for example, touches the number 4 petri dish. Place the entire set of dishes in the incubator for 24 hours at 37 degrees centigrade. (If an incubator is not available, you can place the dishes in a warm place for a similar period of time.)

Hazardous to your health **59**

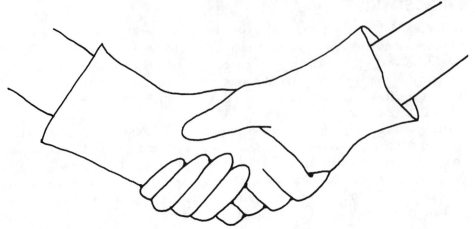

18-2 While wearing surgical gloves, shake hands to see if the microbes can be passed from one person to the next.

Repeat this entire procedure two more times using new gloves and the other sets of petri dishes. After 24 hours have passed, observe the three sets of five dishes for culture growth. Count the number of colonies on each plate and record all the data. The colonies will look like white, shiny, circular mounds on the plate. Use a grease pen to mark the top of the plate above each colony as it is counted. This will prevent you from counting the same colony twice.

Average each number for all three sets of plates. For example, average the number of colonies found in all three of the number 5 plates. Then, compare the averages of the different plates. If one is available, look at the colonies under a dissecting microscope to see their gross structure.

CONCLUSIONS

Were colonies found on all the plates? If so, how many colonies were found on each plate? What was the average for each person's plates? How did they compare with each other? Did they remain constant, diminish or increase as the number increased?

Does it appear that physical contact, such as shaking hands, can transmit microbes from one person to another—even people who never had direct contact with the source of the microbes? If yeasts can be transmitted, is it possible that infectious diseases can do the same?

GOING FURTHER

Research the history of infectious disease epidemics such as the bubonic plague of long ago and AIDS today.

Establish your own disease prevention guidelines for your class or family and then compare them with established public health guidelines.

19
Sugar & starch
Starch digestion

OVERVIEW

Starch is found in many foods. It is composed of glucose molecules (a sugar) strung together in long chains. Animals cannot use the energy stored in starch, but can use energy stored in the smaller glucose molecules. Animals, therefore, must break down the starch into the smaller glucose units to use the energy locked within.

Can the breakdown of starch into smaller sugar molecules be visibly detected? Can the amount of sugar produced be measured? In this project you will mix an enzyme that breaks down starch into glucose and then measure the amount of glucose present. You will use a membrane (dialysis tubing) that allows the glucose to pass through, but not the starch.

MATERIALS

- Two 600 ml beakers
- One smaller beaker
- Unbleached flour from a supermarket or starch from a scientific supply house
- Dialysis tubing (available from a scientific supply house)
- Strong string
- Scissors
- Salivary amylase solution (from a scientific supply house)
- Distilled water or tap water
- Glucose dip stick (available from a pharmacy or supply house)

PROCEDURES

Before beginning, fill one of the larger beakers two-thirds full with distilled water and use the glucose dip stick, according to the accompanying instructions, to test for the presence of glucose. Record your results. Label this beaker "distilled water" and put it aside to be used later.

Next, create a 10% starch suspension by mixing 10 grams of flour or starch with 100 ml of water in the smaller beaker. Label the beaker "10% starch." Then, cut a 6" piece of dialysis tubing and place it in the other 600 ml beaker filled with warm water so it softens. When the tubing is soft, tie off one of the ends with string. Fill the tubing half way with the salivary amylase solution (an enzyme found in your saliva that breaks down starch into glucose). Then, fill the remaining half of the tubing with the 10% starch suspension prepared earlier. The tubing should now be filled. Tie off the top of the tubing tightly with string.

Take the beaker you prepared earlier that is two-thirds full of distilled water and place the filled dialysis tubing (containing the prepared solution) into the beaker. (See Fig. 19-1.) If the tubing is not completely covered, add more water.

19-1 Place the filled tube in the large beaker and immediately take a glucose reading.

Immediately, test the water (outside of the tubing) in the beaker with a glucose stick and record the results. Continue to test the water with the sticks every 10 minutes for one hour, recording the results as you proceed. Plot your results in a graph similar to Fig. 19-2.

CONCLUSIONS

Was there any glucose in the distilled water at the beginning of the experiment? How did the amount of glucose change over time? Why did it change? Explain both the enzymatic process that is occurring and why the glucose moved from the inside of the tubing into the beaker.

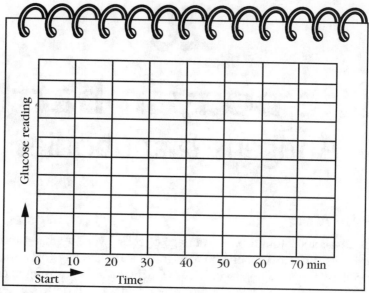

19-2 Create a graph in your journal similar to this one to record your observations.

GOING FURTHER

Study the chemical structure of starch and sugars and the chemical process that occurs during the breakdown of one into the other.

Investigate the number of calories contained within different kinds of foods.

Research why marathon runners eat large amounts of food rich in carbohydrate (*carbohydrate loading*) before a race.

20
The body in battle
Antigens & antibodies

OVERVIEW

Antigens are proteins found on the surface of cells. Every cell has a unique set of antigens on its surface. Antigens on foreign cells are like flags on a ship identifying that cell as foreign. When foreign cells (such as disease causing microbes) enter your body, white blood cells in your blood attack and destroy them.

When the white blood cells attack the foreign cell antigens (enemy flags), the white cells produce antibodies. These antibodies are proteins that clump together the foreign cells so they can be eliminated from your body after they have been destroyed.

People with Type A blood have anti-B antibodies circulating in their blood. Type B individuals have anti-A antibodies. Type AB individuals have neither anti-A nor anti-B antibodies, and type O individuals have both anti-A and anti-B antibodies.

When giving a blood transfusion, it is necessary to find an appropriate donor. What does this mean? Can a type A person receive blood only from another type A individual? Does the blood donor and the recipient necessarily have to be of the same blood type? Can blood be donated or received to and from individuals with different blood types?

MATERIALS

- Type A, B, AB, and O blood samples (about 5 ml of each) (from a scientific supply house)
- Small bottle of anti-A and anti-B antisera (from a scientific supply house)
- Four to six glass slides
- Four to six eyedroppers
- Surgical gloves
- Marker
- Box of wood toothpicks
- Safety glasses

PROCEDURE

Begin by putting on a pair of surgical gloves and safety glasses before conducting this experiment.

Caution: *It is always necessary to take this precaution when handling blood.* (Blood purchased from a scientific supply house, however, should be safe to use for this experiment.)

Place a single drop of anti-B antiserum on each of the four slides. This simulates the anti-B antigens present in type A blood. These slides therefore represent a type A person. Then, mark each of the four slides with one of the following: "A," "B," "AB," and "O."

When the labels are ready, add a drop of type A blood to the slide marked "A" to represent a donation of A blood to the type A person. Add a drop of type B blood to the "B" slide. (*Be sure to use a different dropper for each slide.*) This represents a donation of type B blood to a type A individual. To the "AB" slide add a drop of AB blood to represent AB blood donation. To the "O" slide, add a drop of O blood to represent O blood donation.

Stir the blood gently using a different toothpick on each slide. (See Fig. 20-1.) If the blood forms visible clumps on the slide after a few minutes, the donated blood has been rejected and would be attacked by white blood cells. Make notes of your observations in a table similar to Fig. 20-2.

20-1 Use a toothpick to mix the blood and the antiserum.

20-2 Create a table in your journal similar to this to record your observations.

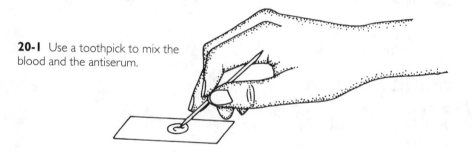

Type "A" individual		(anti-B antisera)		+ = Clumping − = No clumping
A	B	AB	O	Notes
Type "B" individual		(anti-A antisera)		

Clean the slides thoroughly before continuing. Repeat the same procedure by marking the slides in the same way. This time a drop of anti-A antiserum will represent the antibodies found in type B blood. Place a drop of anti-A antiserum on each of the 4 slides. Add drops of the different blood types as previously described on each slide. These represent donations of the various blood types to a type B individual. Stir gently with toothpicks and observe any clumping. Continue to fill in this table.

Clean the slides and repeat once again the instructions stated above, but this time place both anti-A and anti-B antisera on all the slides. This represents the anti-A and B antibodies found in type O blood.

Finally, clean the slides and repeat the instructions one more time by placing a drop of water to represent the blood of a type AB individual, which lacks both anti-A and anti-B antibodies. Add the various blood types, stir and observe any clumping. Complete your notes. Follow the instructions provided by the supply house for the proper method for disposing of the blood after you have completed this experiment.

CONCLUSIONS

From your observations, do blood donors have to be of the same blood type as the recipient? How do you think antibodies affect the compatibility between two blood types? There is a blood type that can be donated to all the other blood types—the universal donor. From your observations which is it: A, B, AB, or O? There is a blood type that can receive all other blood types—the universal recipient. Which one is it?

GOING FURTHER

Read more about antigens, antibodies, and blood types.

Investigate whether blood types unique to humans exist in many types of animals?

Study the different types of blood cells found in different types of animals. How do their form and function vary?

21
Tiny appetites
Ingestion & digestion in a one-celled animal

OVERVIEW

All animals must obtain raw materials to maintain their bodies, grow, and reproduce. This is accomplished by the ingestion and digestion of food. When food is digested, acids are often used to help break down the foods into useable materials and prepare them for absorption into the body.

How does a one-celled animal such as the paramecium "eat" its food and what role does acidity (pH level) play in the digestion process?

MATERIALS

- Active culture of paramecium caudatum (from a scientific supply house)
- Package of bakers or brewers yeast (packaged yeast is usually available in supermarkets)
- Lab spatula
- Beaker (600 ml)
- Sugar
- Pond or tap water
- Graduated cylinder or beaker (100 to 500 ml)
- Hot plate
- Congo Red stain powder (from a scientific supply house)
- Centrifuge or a büchner funnel (possibly available from your school or the funnel can be ordered from a supply house)
- Microscope
- Glass slides and cover slips
- Eye dropper
- Toothpicks

PROCEDURES

Before beginning the experiment, you must prepare an active yeast culture. (The yeast cells will be food for the paramecium.) Do this by adding 50 grams of sugar to 450 ml of tap or pond water in a large beaker. Then, pour the contents of one bag of yeast or a teaspoon of bakers or brewers yeast into the beaker. Stir the liquid and leave the beaker in a warm part of the lab or your home for 24 to 48 hours. (Do not cover the beaker.) After this time, your beaker should contain an active culture of yeast cells.

You must then prepare ("fix") the yeast by killing and staining them. This will allow you to observe the digestion process. To do this, add a spatula full of Congo Red powder to the yeast culture and stir. (See Fig. 21-1.) This will make them more visible and indicate pH at the same time. Place this beaker on a hot plate and boil the culture for 10 minutes to kill and stain the yeast cells.

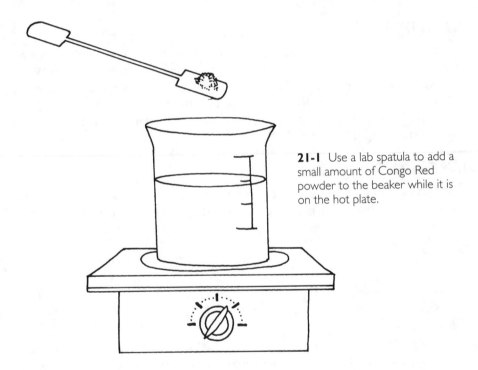

21-1 Use a lab spatula to add a small amount of Congo Red powder to the beaker while it is on the hot plate.

Next, place a sample of the culture in a centrifuge vial and spin to separate the yeast cells from the liquid. (If a centrifuge is unavailable, you can use a büchner funnel that pulls the liquid through a filter by suction.) After using the centrifuge or funnel, pour off the liquid and save the separated, stained, and dead yeast cells found at the bottom of the centrifuge vial. (If you used a funnel, the cells will be on the filter paper of the funnel.) For the moment, leave these fixed yeast cells as is and continue the experiment.

Use an eyedropper to place a small drop of the paramecium culture on a microscope slide and cover with a cover slip. Examine the organism using first the low power objective and then a higher power. Take notes and draw illustrations.

After observing their normal behavior, remove the cover slip and dip a toothpick into the fixed yeast cells prepared earlier (now in the centrifuge vial or on the filter paper). Stir the toothpick onto the slide and replace the cover slip. Again, observe under a high power. Study how the paramecium ingests the yeast cells. (See Fig. 21-2.) Draw illustrations of what you see. Continue to observe when the yeast cells are in the paramecium's cytoplasm and note any color changes that occur. (This process should take about 20 minutes.)

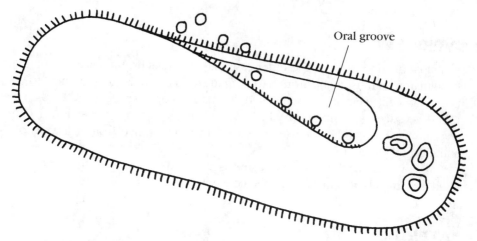

21-2 Food enters the body of the one-celled paramecium through the oral groove.

CONCLUSIONS

How does the paramecium ingest the yeast cells? What happens to the cells once they are inside the paramecium? Since Congo Red changes colors when the pH changes, research what is happening to the yeast cells as they are digested by the paramecium.

GOING FURTHER

Investigate how acids are used by the paramecium to digest the food. Relate the digestion of this single celled organism to humans.

Perform a comparative anatomy study of how animals bring food into their bodies. Start with single-celled animals and go up the evolutionary ladder.

22
Getting there
Osmosis & diffusion

OVERVIEW

Cells must obtain nutrients and other essential materials from their environment. These materials must pass through the organism's cell walls (membranes) to get inside the cell. This is accomplished in many ways. These include passive processes, such as osmosis, in which water passes from higher to lower concentrations, and active processes in which materials move from a lower concentration to a higher.

Can you observe the process of osmosis in which water passes through a simulated cell membrane from a higher concentration (100% distilled water) to a lower concentration (glucose solutions containing a smaller percentage of water)?

MATERIALS

- Dialysis tubing that will simulate a cell membrane (available from a scientific supply house)
- Thistle tube or pipette
- Stand and clamp
- Sugar
- Graduated cylinder (100 ml)
- Scale or balance
- 600 ml beaker
- Three 150 ml beakers
- Scissors
- Strong string
- Glass marking pencil
- Ruler

PROCEDURE

First, prepare a 5% and a 25% sugar solution by mixing 5 g (grams) of sugar with 100 ml of tap water in a beaker. Label this beaker "5% solution." Then, prepare a 25% solution by mixing 25 gr of sugar and 100 ml of tap water in another beaker and stir. Label this beaker "25% solution." Fill a third beaker to the same level as the others with plain tap water and label it "control."

Next, prepare the stand and clamp to accept the thistle tube or pipette as shown in Fig. 22-1. Then, cut a 6" strip of dialysis tubing and place it into a 600 ml beaker of warm water to soften. After it becomes soft, tie off one end of the tube with a string. Pour the 5% glucose solution into the open end of the tubing. Attach this open end to the thistle tube as shown in Fig. 22-1 and tie it on securely.

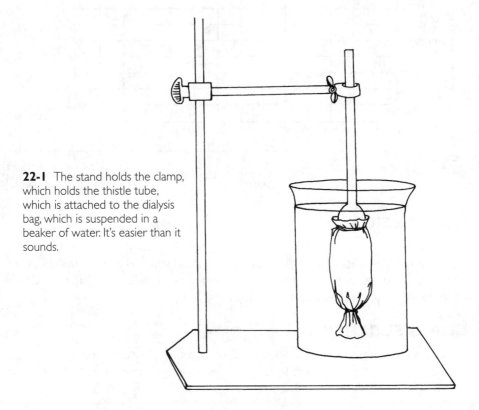

22-1 The stand holds the clamp, which holds the thistle tube, which is attached to the dialysis bag, which is suspended in a beaker of water. It's easier than it sounds.

Attach the thistle tube and the attached dialysis tubing to the clamp and position it so the filled tubing is submerged into a beaker of water as shown in Fig. 22-1. Immediately use the marking pencil to mark the level of the solution at the top of the thistle tube. Record the distance from the top of the thistle tube to the mark. Continue to mark this level every 10 minutes for two hours. Plot the data in a graph as you see in Fig. 22-2.

Once this is complete, repeat the entire procedure stated above using the 25% sugar solution. For the last group, repeat the procedure, but use pure tap water in the dialysis tube instead of a sugar solution. This will be the control group. Plot the data for these two groups on the same graph.

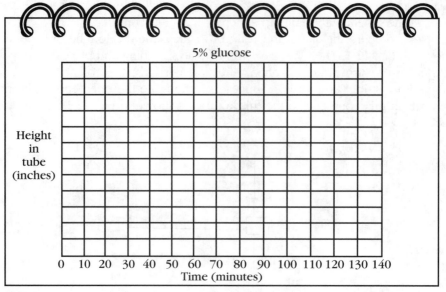

5% glucose

Height in tube (inches)

0   10   20   30   40   50   60   70   80   90   100   110   120   130   140
Time (minutes)

22-2 Plot the data in a graph similar to this one for the 5% and 25% solutions and for the control.

CONCLUSIONS

What happened to the level of fluid within the thistle tube over time? Did it go up or down? Why did it go in the direction it did? How did the results differ between the two experimental groups and the control? Why was there a difference? How does this relate to the cells in our bodies?

GOING FURTHER

Create an experiment that studies the effect of temperature on this experiment.

Research why physiological saline is used to wash out wounds and to clean contact lenses instead of tap water?

Part five

Animal lives

Do all spiders of the same species create similar webs? Is a larger spider likely to make a larger web? These questions are investigated in the first project in this section. The second project tries to explain the common phenomenon of earthworms surfacing after a rainfall. How wet does the soil have to be to make them surface; moist, damp, soaked, saturated?

The third project has you first perform a literature search to find information about the gestation periods, life spans, and number of offspring produced by a wide range of types of animals. It then has you analyze the data you've collected to look for relationships among these three factors. This is a perfect project to perform on a computer.

The fourth project uses a goldfish to study the effect of water temperature on respiration rate. Respiration rate is an excellent indicator of the overall metabolic activity of an animal. The final project in this section has you determine those conditions most important to the survival of an animal. You'll do this by looking for conditions that are always present in areas where the animal is found. You'll learn about tolerance ranges, optimum levels, and limiting factors.

23
Trapped
Spiders & webs

OVERVIEW

Some of most beautiful natural works of art are created by spiders—their webs. Their beauty, however, doesn't tell the whole story. They spin these silky webs to trap unsuspecting prey so they can suck the life out of them.

Are these webs like snowflakes, with every one different than the next, or are they all similar? Do members of the same species create similar webs? In addition to these questions, you can also investigate whether the size of a spider affects the size of its web.

MATERIALS

- At least a dozen large sheets of black poster board
- Talcum powder
- Stick glue (the kind that can be easily spread on paper)
- Access to many spider webs
- Scissors
- Ruler
- Spider identification guide
- Gloves

PROCEDURES

During the warm season, go on a field trip in search of spider webs. The best place to look is in open fields with tall grasses and vegetation. To find other kinds of webs, look in the woods or the tall grasses around ponds.

When you find a web, record its location. Take measurements of the web. (See Fig. 23-1.) Look for the spider—it usually lies in wait around the edge of the web.

Caution: *Wear gloves and be careful around spiders because they are poisonous and some can bite.*

Use your field guide to identify the spider and use the ruler to measure the length of the spider. (You can collect spiders using forceps and then place them

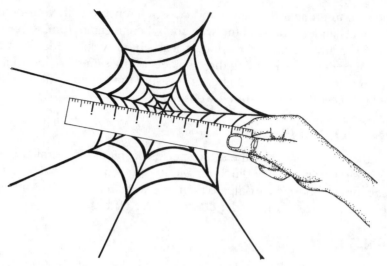

23-1 Measure the width of the spider web in its natural state.

in a vial with 70% alcohol, but this is not required unless you want to make a collection for this project.)

Next, you will collect the web. To do this, remove the spider from the web with the forceps and place it on another plant. (It will probably leave on its own from all the commotion.) Gently dust the web with the talcum powder so there is a light coating. Cut a piece of black poster board a little larger than the size of the web. Apply a thin layer of the stick glue over the entire poster board.

Then, hold the poster board up to the web as you see in Fig. 23-2 and gently touch the board to the web so it sticks. Have your partner look for the strands of the web that connect it to the vegetation and cut them, freeing the web.

23-2 Touch the poster board, now covered with glue, to the web and have a friend cut the main supporting strands that hold it in place.

Be sure to number the board and cross reference it with your notes so you know which spider produced the web and where it came from. Repeat this procedure for many webs. Try to find at least three webs made from the same species of spider. When finding three spiders of the same species, look for individuals of varying size. (Remember to measure the length of all the spiders.) If possible, collect five sets of three webs.

CONCLUSIONS

Did all the webs created by the same species of spider look similar in their geometric design, or were they different? Did the webs of different species have different geometric designs? Compare the three webs made by the same species of spider. Did the size of the spider effect the size of the web?

GOING FURTHER

Investigate how a spider builds its web. Does it start from the outside and work in, or start from the inside and proceed out?

Create a project to determine whether different types of webs are used to trap different kinds of prey.

Create a project to test the tensile strength of the varying types of silk used in webs.

24
I'm drowning
Water & earthworms

OVERVIEW

We've all seen earthworms come to the surface after a rainfall. Earthworms need air to survive, just as we do. They use air that is trapped in the spaces found in the soil. If these air spaces become filled with water, the worms must come to the surface. How much moisture does it take to force worms to surface? Does the soil just have to be moist or must it be soaked to force worms to the surface?

MATERIALS

- About one dozen earthworms (can be collected or purchased from a bait shop, garden center, or a scientific supply house)
- Soil (preferably from a garden or the woods)
- Two plastic or rubber storage containers that hold a known volume (a 5 quart tub that is about 5" deep, 8" wide, and 13" long would work well)
- Potting soil
- Distilled water (tap water is okay)
- Water bottle sprayer or mister with graduations to measure the amount of water in the bottle
- Spoon
- Small, narrow ruler

PROCEDURES

Prepare both tubs by filling them with 6" of soil—an inch from the top rim. Don't pack the soil. (You will be using two tubs to verify data collection.) Place six worms in each tub and let the worms settle in for one hour. Most of the worms will burrow into the soil.

Fill the water mister and record the level in the mister bottle. Note whether any of the worms are at the surface before you begin. Spray water evenly over the surface of each tank. (See Fig. 24-1.) Just moisten the surfaces; don't flood them. Wait 10 minutes and observe whether any worms surface. Measure the amount of water dispensed over the surface. Record your actions and observations.

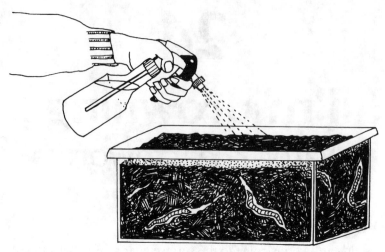

24-1 Spray the soil evenly and measure the amount of water used.

Spray the entire surface in both tubs once again with another, but heavier mist. Record the total amount of water used and wait 10 more minutes to see if the worms surface. Every 10 minutes, increase the amount of moisture applied, record the total amount of water used, and wait to see if the worms surface.

As you proceed, record the general appearance of the soil after each application. Fill in a table similar to Figure 24-2. Is the soil moist, wet, soaked, puddles, etc? Continue this procedure with both tubs until the worms begin to surface. Once they begin to surface, continue applying only a light misting every 10 minutes until they have all surfaced.

Spray #	Total volume sprayed	Number of worms at the surface	Condition of soil
1	ml	0	Slightly moist
2	ml	0	Slightly moist
3	ml	0	Damp
4	ml	3	Wet
5	ml	5	Saturated
6	ml	6	Saturated to 2 inches
7	ml		
8	ml		
9	ml		
10	ml		

24-2 Create a table in your journal similar to this to record your observations.

Once all the worms have surfaced, remove them and take a final reading of the total amount of water used. Then, determine how much of the soil is wet. Use a spoon to dig down into the soil to see how deeply it is wet, or whether it is saturated. Use a ruler to measure the depth of the wet soil.

CONCLUSIONS

How much water was needed for how much soil to bring the worms to the surface? Will worms surface because the soil is simply moist, or damp, or must the soil be saturated? If it must be saturated, how much (how deeply) must it be saturated?

GOING FURTHER

Run a similar experiment, but use different types of soils that can hold different amounts of air and water.

Run a similar experiment, but use different depth tubs to determine if worms first dive deeper to avoid water or move toward the surface immediately.

25
Pregnancy, birth, & longevity
Relationships

OVERVIEW

The term *gestation* refers to the length of time a female animal remains pregnant. An animal's ability to produce many offspring (babies) is called *fecundity*. *Longevity* (life span) refers to how long an animal lives.

In the first part of this project, you will do a literature search to find variations in typical gestation periods, numbers of offspring produced, and life spans throughout the animal kingdom. What are the shortest and longest life spans of all the animals species? What is the longest and shortest gestation period of all the species? Which species of animal produces the smallest and which the largest number of offspring?

Once you've concluded search, you'll study this data to look for relationships that exist. For example, is there a relationship between gestation and longevity, or between longevity and fecundity?

MATERIALS

- Access to a good library containing extensive information on the biological sciences
- Computer (optional)

PROCEDURES

Create a table similar to Fig. 25-1. It would be best if you can create this table on a computer, using a spreadsheet or word processing program, since you will continually be adding new information as the research progresses. (The computer can also be used to graph the data, which will help you analyze it.)

First, begin collecting the data (gestation period, longevity, and fecundity) for as many animals as possible, to be sure to survey the entire animal kingdom. Begin with simple organisms (invertebrates) and then move on to the higher

Animal	Gestation period	Typical # of offspring per female	Typical lifespan	Misc. notes
1.				
2.				
3.				
4.				
5.				
6.				

25-1 During your literature search, fill in a table similar to this one. It would be best if you could create this table on a computer.

forms of life (vertebrates). The more animals included in the table, the better. For the gestation period and longevity, enter the hours, days, months, or years. For fecundity, enter the number of offspring that a female typically produces during her life time.

After you have completed collecting all the data, it is time to analyze it. Use the data to draw or create graphs on the computer. If you are using a computer graphing program, you can design interesting graphs to represent all the data simultaneously. This will be the best way to look for relationships. For example, plot all three figures on a single graph for each type of animal. (You will use both the left and right Y axes to represent two of the figures, plus your imagination to display the third.) Print out on a transparency of all of these graphs for each animal. Then you can overlay each transparency on the others to look for relationships.

CONCLUSIONS

Are there large differences found in the gestation periods, longevity, and fecundity throughout the animal kingdom? Do there appear to be any relationships existing among any or all of these factors?

GOING FURTHER

Add another factor to the table, such as the animal's size. Does this appear to be related to the other factors in any way?

Research *K-strategists* and *R-strategists*. How do these terms relate to this project?

26
Heavy breathing
Respiration rates & fish

OVERVIEW

The term *metabolism* refers to all the activity going on in a living organism. This includes physical activities, such as movement, and chemical activities, such as respiration. Most of these activities are closely associated to temperature. In warm-blooded (*endothermic*) animals, the body temperature remains relatively constant regardless of the environmental temperature and the rate of chemical reactions within their bodies also remains constant.

In cold-blooded (*exothermic*) animals however, the body temperature is determined by the temperature of their environment and can fluctuate wildly. Does the metabolic activity of exothermic animals change as the temperature of their environment changes?

In this project, you will study the respiration rate of a goldfish in waters of varying temperatures. Will the respiration rate indicate a reduction in the metabolic rate?

MATERIALS

- Two or three goldfish
- Aquarium net
- Thermometer
- Two small bowls for each goldfish (quart-sized, wide mouth jars are fine for most small goldfish)
- Bucket of crushed ice

PROCEDURES

Place one goldfish in each jar. Leave them undisturbed for 5 minutes in room temperature water. (Use a thermometer to determine the actual temperature.) Then, begin observing the movement of their gill covers, called *opercula*. (See Fig. 26-1.) Be sure not to disturb and excite the fish. Count the number of times the gill covers open each minute. Do this three times for each of the three fish.

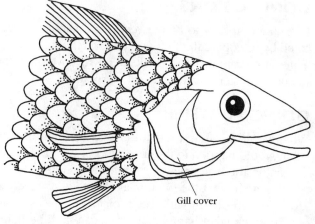

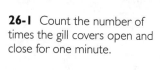

26-1 Count the number of times the gill covers open and close for one minute.

Gill cover

Prepare another set of three jars, but cool the water using the ice so it is about 15 degrees F cooler than the first bowls. Use the thermometer to read the temperature. Use the net to transfer the fish into the cooler jars. Leave the fish undisturbed for five minutes and then begin counting the gill cover movement once again. Add ice if necessary to maintain the proper temperature and repeat your count for this fish. Repeat this procedure for the other fish.

Repeat this procedure, but this time place the fish in jars cooled to just above the freezing mark. (You can use the first "room temperature" jars for this.) Transfer each fish to one of these jars and once again count the gill movement for one minute. After recording all the data, plot a graph that illustrates this data. (See Fig. 26-2.) Return the goldfish to the tank when the experiment is finished.

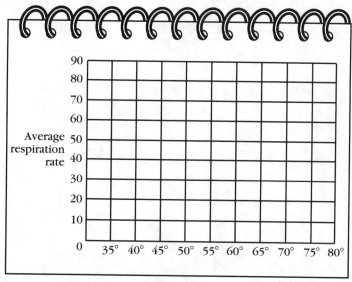

26-2 Plot your data in a graph similar to this one.

CONCLUSIONS

Does there appear to be a direct relationship between the respiration rate indicated by the operculum movement and the water temperature? Does cooling the water appear to increase or decrease the overall metabolic rate of the fish? Why would water temperature affect respiration?

GOING FURTHER

What would happen if the water was warmed instead of cooled in this project? What would happen to the respiration rate? Do factors such as dissolved oxygen content in the water play a role in the results?

Research what the beneficial effects of a "fever" are on humans who are ill? How does this tie in with this project?

Research what happens to animals that hibernate or estivate and relate your findings to this project.

27
Home sweet home
Surveying an animal's preferred habitat

OVERVIEW

Animals must live in an environment that meets their needs. They must have the correct amount and kinds of nutrients available to eat and oxygen to breath. Aquatic animals must live in water containing the correct amount of salts and dissolved oxygen.

For an animal to survive, all of these conditions (nutrients, oxygen, salts, etc.) must fall within the minimum and maximum levels needed by that animal. If a condition such as temperature goes above or below the limits, it becomes the *limiting factor*. This might be below freezing and above 90 degrees F. This range of any given factor within which an animal can survive is called the *tolerance range*.

If a condition is perfect for an animal to survive, it is said to be at the *optimum level*. For example, a water temperature of 77 degrees F might be the perfect temperature for a certain kind of protozoan or fish, meaning it is at the optimum level. Animals will always seek out an environment that has all the conditions necessary to survive within the proper tolerance ranges and as close to the optimum level as possible.

In this project, you'll try to determine some of the conditions that are important to an animal's survival by surveying all the habitats where they live. You'll look for conditions that are always present.

MATERIALS

- Access to many different habitats (for example, woods, open fields with vegetation, dry areas, shaded areas, etc.)
- Field notebook

PROCEDURES

First, select the kind of animal you want to survey. A few good ones to begin with are earthworms, pill bugs (affectionately called *roly-polies*), or slugs. Check an identification guide to be sure you can identify the animal that you have selected.

Have an adult accompany you on your field trips and bring a field note-book. You want to search for the selected animal in as many places (habitats) as possible and take detailed notes about where you did and did not find them.

The most important part of this project is to record the environmental conditions of each habitat. If you search an area that is very dry and usually in the direct sunlight, and find none of the selected type of animals, record this information. If you always find them in habitats that are in shade and damp areas, this should become obvious in your notes. (See Fig. 27-1.) Include the numbers of individuals found in each area, as well. The more areas you search, and the more diverse these areas, the better.

Pill bugs		
Area searched	Enviromental conditions	#'s found
1. North side of house	Shady, damp soil, covered with fallen leaves	3
2. South side of house	Sunny, gravel driveway, no vegetation, dry	0
3.		

27-1 Create a table in your journal similar to this to record your observations.

CONCLUSIONS

Review your notes and look for similarities among all the areas that contained the animals. What conditions kept appearing in your notes whenever the selected animal was found and which kept appearing when they were not found? Does soil, moisture, sunlight, or some other condition appear to be directly related to the presence or absence of the animals?

Once you've selected a condition that appears to be important to the animal, see if you can determine the optimum level of that factor. For example, the largest number of individuals were found in areas that you categorized as

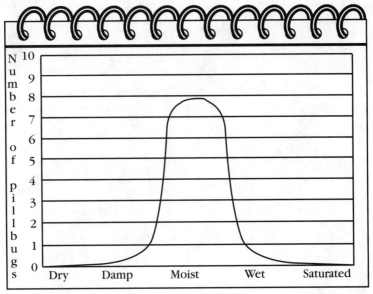

27-2 Plot your data in a graph similar to this one.

"damp" as opposed to "wet," "soaked," or "dry." Draw a graph that shows the numbers of individuals found in each type of habitat searched (Fig. 27-2).

GOING FURTHER

Try to confirm your results from the field by reproducing them in the lab. Create an experiment that gives the animal a choice among different conditions such as dry, moist, damp, and soaked soils to see which they prefer. Do the results match your field studies?

Study how environmental contaminants affect where an organism can survive. How does this relate to the project?

Part six

Communications & senses

Animals communicate in many ways, but none are as interesting as *bioluminescence*. In the first project, you'll use a penlight to try to communicate with fireflies and learn how they communicate with each other. The second project investigates human sensory perception by looking into the role vision plays in our sense of balance.

The third project in this section is also about sensory perception, but in invertebrate animals such as millipedes, earthworms, and insects. Do these animals have specialized organs to sense "smell" or do they use a simple net of receptors that cover their bodies?

We've all seen night-flying insects swarming around a porch light on a warm summer night. The fourth project has you determine whether these critters are being attracted to the light or the heat created by the bulbs. The last project in this section has you determine how the sense of smell plays a role in your sense of taste.

28

You light up my life
Communicating with fireflies

OVERVIEW

Bioluminescence (light produced by living organisms) is used by many animals to locate and attract a mate. In some cases, it is the shape of the light that attracts others. In other cases, it is the sequence of the flashing lights that attracts the mate.

Fireflies, which are actually beetles, use bioluminescence to locate a mate by a certain sequence of flashes. (See Fig. 28-1.) In most species, the female remains on the ground flashing her light to attract males. When males of the same species see the flashes, they respond with their own flashes until they meet.

Different species of fireflies use different flash sequences. This allows individuals to locate only members of their own species. Is it possible to get fireflies to respond to artificial flashing lights (flashlight) by imitating their flashing sequence? Can you determine their flashing sequence by their response or lack of it?

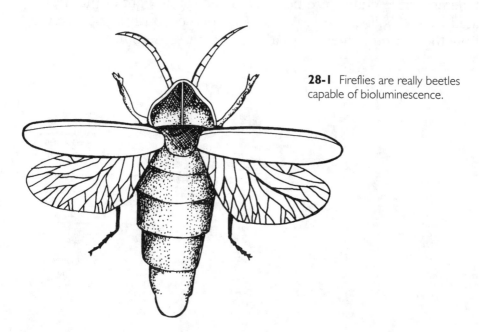

28-1 Fireflies are really beetles capable of bioluminescence.

MATERIALS

- Access to fireflies (try to catch them all in the same location at the same time—you'll need at least 10)
- Jar with a lid or some cheesecloth to cover the jar and rubber bands to hold the cheesecloth in place
- Small narrow flashlight (preferably a penlight)

PROCEDURES

Collect the fireflies and place them in the jar. Put the lid on the jar loosely or cover with cheesecloth. Bring the jar into a dark room. Observe the lighting patterns and record your observations. (If you must turn on the flashlight to write, do this under the table so it doesn't interfere with the project.)

Remove one firefly and put in it a separate jar. This firefly will be the test subject. Hold the flashlight about one foot from the jar and flash the light on for one second and then off for one second. Repeat this sequence for 30 seconds. Notice whether the insects appear to respond (flashback to you) in any kind of pattern. Record your observations.

Repeat this procedure, but change the sequence to one second on and 2 seconds off. Repeat this sequence for 45 seconds. Repeat the above procedures for one second on and 3, 4, and 5 seconds off, recording the results. (Increase the time period so you always turn the penlight on 15 times.) Repeat this process with at least ten fireflies. Release the fireflies when you complete this experiment.

CONCLUSIONS

Did the fireflies respond to your flashing? If so, which sequence did they respond to the most? How did they respond? What was the sequence and timing of their flashing? Did the response sequence match yours to some degree?

GOING FURTHER

Study the mating behavior of fireflies.

Research how bioluminescence is used in other animals.

Research how bioluminescence works biochemically.

29
Standing
with your eyes
Balance & vision

OVERVIEW

We can stand upright without falling on our face because, in part, we have a sense of balance. The primary balancing organ is found in our ear and is called the *semicircular canal*. When a person has an ear infection, they might feel disoriented or dizzy, since their sense of balance is affected by the infection. But is this the only organ involved in balance? Do our eyes play a role in our sense of balance?

MATERIALS

- Stopwatch or a watch with a second hand
- Blindfold (a handkerchief will work fine)
- At least four friends to help you

PROCEDURES

Have a friend use the stopwatch to measure how long you can stand on one foot with your eyes open. Take turns with the watch so that all five participants are timed standing on one foot with eyes open. Record the data for each person in a table similar to Fig. 29-1.

Repeat this entire procedure two more times for each person with their eyes open. Then, repeat the entire procedure for each person, but have them stand on one foot while blindfolded, or with their eyes closed. Again, take notes and fill in the table. Repeat this procedure two more times with eyes covered or closed. Then average all the data for each person and finally, take the average for all the people with their eyes open and their eyes closed.

| Name | Trials | | | | | | | |
| | Eyes open | | | | Eyes closed | | | |
	#1	#2	#3	Avg.	#1	#2	#3	Avg.
John								
Joe								
Mary								
Fred								
Group avgs.								

29-1 Create a table in your journal similar to this to record your observations.

CONCLUSIONS

What was the average for all the people with their eyes open and closed? Was there a big difference? Does it appear obvious that the eyes have some role in balance?

GOING FURTHER

Research the role of the eyes in balance. How does it work?

Develop other ways to test how our eyes affect balance.

30
It stinks
A survey of sensory perception

OVERVIEW

We smell with our nose—a highly specialized sensory organ. Other organisms sense smell with a variety of organs or simple receptor cells. Do the earthworm, millipede, snail, and grasshopper have specific localized organs (like our nose) that detect smell, or is there a disperse network of receptors (like touch receptors in our skin) that detect smell in these animals?

In this experiment you'll test the response to a strong smell on different parts of each of the above animals to see what part of their body "smells" a substance.

MATERIALS

- A few earthworms
- A few millipedes
- A few slugs or snails
- A few crawling insects
- Cotton balls
- Nail polish remover
- Coated paper plates

PROCEDURES

 Place the first millipede on the paper plate. Pour a little nail polish remover on a cotton ball and place it close to the millipede's head as you see in Fig. 30-1. (*Don't inhale this nail polish remover.*) Don't actually touch the cotton ball to the millipede. Record the response, if any. Look for the animal to jerk or move away from the cotton ball. If there is no response, record this as well.

Next, place the cotton ball near the posterior end and record the response, if any. Finally, place it near the middle and record the response. Repeat this procedure for each of the three individuals.

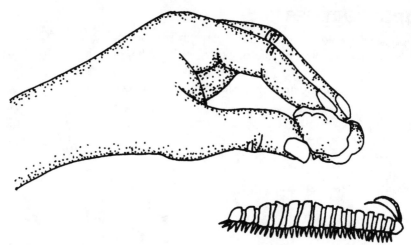

30-1 Hold the cotton near the head of the millipede but don't touch it.

When you have finished using the millipedes, release them and continue the experiment by following the same procedures for the slug, earthworm, and insects. Record all of your observations in a table similar to Fig. 30-2.

Animal	Response (+ means the animal moved away from its cotton)			Notes
	Head	Mid-section	Rear	
1. Earthworm				
2. Grasshopper				
3. Millipede				
4.				
5.				
6.				

30-2 Create a table in your journal similar to this to record your observations.

CONCLUSIONS

Did any or all of the animals respond to the offensive smell in all three positions? Did any respond only when it was placed near the head? Research the location of the sensory receptors in all of these animals and determine why some responded to all three areas and others to only one.

GOING FURTHER

Investigate the role "segmentation" plays in sensory perception. (Consult an invertebrate anatomy and physiology text book.) What does it have to do with this project?

Study the anatomy and physiology of sensory reception.

31
Searching
in the dark
What attracts
night-flying insects?

OVERVIEW

Some animals are attracted to heat, while others are attracted to light. You've probably seen a swarm of moths flying around a porch light during the warm summer months. Are these insects attracted to the light being given off, by the heat generated by the bulb, or both? In this project you'll try to determine whether it is the heat, light, or a combination of both that attracts night-flying insects.

MATERIALS

- Fluorescent bulb and fixture
- Similar-rated incandescent bulb and fixture
- Hot plate (from your school lab)
- Heavy-duty extension cord for outdoor use
- Heavy table to support the lamps and hot plate

PROCEDURES

Warning: *Perform this experiment on a dry night when there is no chance of rain.*
    This project should be done during the warm summer months and with a full moon so you can see what is going on. Set the table up outside. (The extension cord must be able to reach the table.) Have your sponsor, teacher, or parent prepare all the electrical equipment including the extension cord for you. While using these electrical devices be cautious.
    After dark, first prepare the hot plate (which produces no light, but a lot of heat) by placing it on the table and plugging it into the extension cord. (See Fig. 31-1.) Turn the hot plate on to high. *Be sure it does not present a fire hazard.*

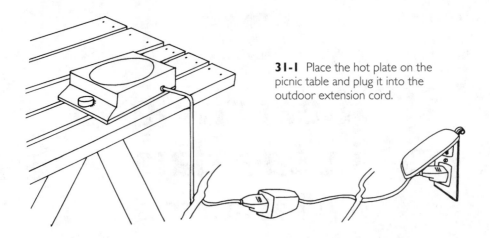

31-1 Place the hot plate on the picnic table and plug it into the outdoor extension cord.

Quietly stand back and observe for 30 minutes what kinds and how many visitors appear around the hot plate. Record your observations, including the types and numbers of visitors. (A general description is all that is needed. For example, three moths, and two lacewings.)

Then, turn off the hot plate, wait until it cools and remove it. Now, hook up the fluorescent bulb and fixture (which produces a lot of light and very little heat). Place it on the table facing up. Again, observe the visitors for 30 minutes and record the data. (See Fig. 31-2.) Finally, repeat the procedure using the incandescent bulb and fixture (which produces light and heat). If possible, repeat the entire procedure two or three more times in different locations and compare the results.

Device	Kinds & numbers of visitors		Notes
1. Hot plate (heat)	a. _____ b. _____ c. _____	d. _____ e. _____ f. _____	
2. Fluorescent (light)	a. _____ b. _____ c. _____	d. _____ e. _____ f. _____	
2. Incandescent (heat & light)	a. _____ b. _____ c. _____	d. _____ e. _____ f. _____	

31-2 Create a table in your journal similar to this to record your observations.

CONCLUSIONS

What kinds of organisms and how many of each were attracted to each device? Which attracted the most visitors? Does it appear that insects are attracted primarily to just heat, just light, or both?

GOING FURTHER

Research why these organisms are attracted to what your results showed.

Continue this project to see if a moving light will attract insects as well as a stationary light.

Research how fireflies use the light they produce. Are they using the light or the heat? How efficiently do they produce this light compared to our light bulbs?

32
Tasting with your nose
Smell & taste

OVERVIEW

When you think about taste, your tongue most likely comes to mind. Your tongue has taste receptors that detect sweet, bitter, sour, and salty foods. These receptors are dispersed over the surface of the tongue in certain regions.

But what role, if any, does smell play in taste perception? Can you still "taste" food without the ability to smell that food? Is smell important to taste many different types of foods or only a few?

MATERIALS

- Apple, pear, onion, and a few other fruits or vegetables of similar texture
- Handkerchief to act as a blindfold
- Nose clip (the kind swimmers use)
- Forks or toothpicks
- Two or three friends to help

PROCEDURES

Do not allow your friends to know the types of foods being prepared. Cut each of the foods into bite-sized pieces. Place the blindfold and nose clips on a friend and bring the participant into the room. Place the first food morsel on the fork and hand it to your friend to taste. Give them 10 seconds to determine what it is. Record their first guess. Repeat this procedure with each different type of food.

After the person has tried each type of food with the nose clip, repeat the procedure without the nose clip. Be sure to mix up the sequence in which foods are offered. Record their accuracy once again. Repeat this entire procedure with the other participants. Fill in a table similar to Fig. 32-1.

| Person tested With noseclip | Foods offered | | | | | % correct | Notes |
	Correct = Y			Incorrect = N			
	#1	#2	#3	#4	#5		
1. Lindsay							
2. Kim							
3. Nancy							
Without noseclip							
1. Lindsay							
2. Kim							
3. Nancy							

32-1 Create a table in your journal similar to this to record your observations.

CONCLUSIONS

How many of the foods did each person get correct with the nose clip and without the nose clip? Of all those tested, what percentage did they get correct while they could and could not smell? Was there a significant difference? How important does smell appear to be in tasting your foods?

GOING FURTHER

Research how some animals have organs, such as antennae, that respond to many different stimuli such as taste, smell, and even touch.

Research the comparative anatomy and physiology of smell in a variety of animals such as dogs, reptiles, and insects.

Part seven

Growth & development

Have you ever been told that it is healthier to eat raw fruits and vegetables than cooked ones? Many people believe that raw foods contain more vitamins and minerals than cooked ones. The first project in this section investigates whether this belief is true.

Sex ratio refers to the number of males to the number of females in a population. The second project has you determine if a change in temperature changes the sex ratio in a population of mosquitoes. Will cold weather produce more males or females? The third project returns to investigating human growth. How often do you cut your fingernails? Do fingernails grow faster in some groups of people? Would yours grow faster if you took vitamin pills regularly?

In the fourth project in this section, you'll see if an insect's egg or pupal case is designed to withstand the freezing temperatures of a northern winter. The fifth project uses the scales of fish to help you determine their age. Can we count marks on fish scales just like the rings of a tree trunk to determine age? The final project investigates regeneration in simpler forms of life including the planarian and the starfish. Can you cut a planarian into any two pieces and get two new individuals? How about starfish?

33

Vitamins & cooking
Is raw food more nutritious than cooked food?

OVERVIEW

Humans have a rather odd habit when it comes to eating their vegetables—they cook them. Wild animals do fine by simply eating their vegetables (and all other foods) raw. One advantage of cooking foods is that it helps destroy any unwanted disease-causing microbes that might be living on the foods.

It is often said that cooked foods are less nutritious than raw foods because cooked foods lose their vitamins and minerals. Is this more a myth than a reality? Does cooking really destroy vitamins?

MATERIALS

- Raw, fresh cabbage leaves
- Food blender
- Burette (50 ml) (available at your school or from a scientific supply house)
- Burette clamp and stand (same as above)
- Funnel
- Funnel support and stand
- Filter paper
- 250 ml beaker
- Two 50 ml beakers
- 0.2% indophenol solution (from your school or a scientific supply house)
- 10 ml graduated cylinder
- 500 ml graduated cylinder
- Hot plate

PROCEDURES

Place five or six large raw cabbage leaves into a blender with about 200 ml of water and blend until smooth. Place the filter paper into the funnel and the funnel over a 250 ml beaker. Pour the contents of the blender into the funnel. (See Fig. 33-1.) Filter out and save the filtrate (juice) in the beaker. Label this beaker "filtrate."

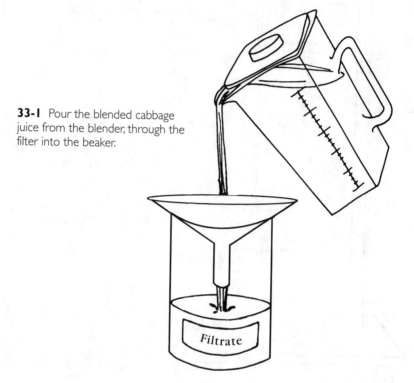

33-1 Pour the blended cabbage juice from the blender, through the filter into the beaker.

Place the burette into the clamp so it is ready for use as you see in Fig. 33-2. Pour 10 ml of the juice from the "filtrate" beaker into the top of the burette. Record the volume in the burette before proceeding. Pour 10 ml of the indophenol into a 50 ml beaker and place this beaker under the burette as you see in Fig. 33-2.

Titrate the juice into the indophenol beaker. This is done by turning the stopcock and allowing the juice in the burette to drop down into the beaker below. Allow the titration (dripping) to continue until there is a sudden change in the color of the beaker from blue to clear. (In some instances, it might appear yellowish in color.)

As soon as the color change occurs, note the number of milliliters remaining in the burette. Subtract this number from the original volume (10 ml) to determine how much juice was needed to change the color of the beaker containing the uncooked cabbage solution. The amount of juice needed is an in-

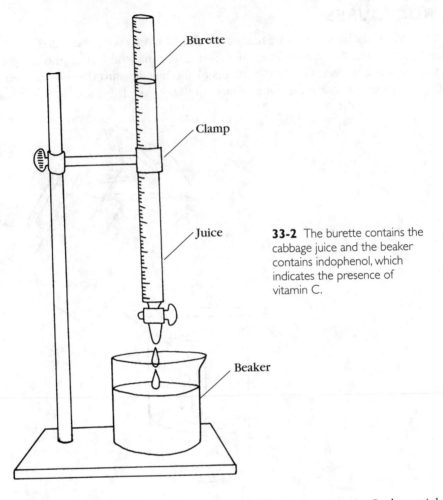

Burette

Clamp

Juice

33-2 The burette contains the cabbage juice and the beaker contains indophenol, which indicates the presence of vitamin C.

Beaker

dication of the amount of vitamin C present. The more vitamin C, the quicker the solution will change color. You now have a baseline for the amount of vitamin C present in raw cabbage.

Next, you'll determine if the same amount of vitamin C is present in cooked cabbage. Boil the remaining cabbage juice (stored in the beaker labeled "filtrate") for 20 minutes over a hot plate and then allow it to cool. Repeat the steps above by filling the burette with 10 ml of this cooked juice, placing 10 ml of indophenol in a 50 ml beaker, and titrating until the color changes. Again, note the amount of juice needed to change the color. Record your observations.

CONCLUSIONS

Plot your data on a graph to visualize the results. How much raw juice was needed (titrated) to change the color of the indophenol? How much cooked juice was needed? (The more juice needed to change the color, the less vitamin

C is present in the juice.) Does cooking cabbage appear to reduce the amount of vitamin C present in the cabbage? Are you better off eating raw or cooked cabbage?

GOING FURTHER

Explore how different methods of preserving foods affects their nutritional value. For example, how do canned, frozen, and fresh vegetables differ in vitamin content? You can do this both by performing experiments and by researching the labels found on food packaging.

Run a similar experiment with other types of foods. Discuss with your sponsor which foods to use.

34
Boys & girls
Sex ratio & temperature

OVERVIEW

The number of males to females produced in a population is called the *sex ratio*. In most populations the ratio is close to 50:50, meaning for every 100 offspring there will be 50 males and 50 females. There are many exceptions to this rule, however. In some instances, the sex ratio varies. This may be due to changes in season, predator-prey relationships, changes in weather, and other factors.

This experiment attempts to determine if the sex ratio of mosquitoes is affected by changes in temperature.

MATERIALS

- Access to about 150 mosquito eggs or larvae often called *wrigglers* (these can easily be found during the warm months or the eggs can be purchased from a supply house)
- Seven similar, quart-sized jars with screw caps
- Enough cheesecloth to make covers for all the jars
- Large rubber bands to hold the cloth on the mouths of the jars
- Small brush (a model plane paint brush works well)
- Small aquarium fish net to collect the wrigglers
- Killing jar (available from a scientific supply house)
- Incubator (or a warm place in your home or school that maintains a temperature of about 90 degrees F such as a boiler room)
- Dissecting microscope or magnifying glass to determine the sex of the adult mosquitoes

PROCEDURES

 Mosquito larvae (wrigglers) or eggs (usually found in groups or rafts) can be found in still waters such as a small pond, a bird watering tray, or long standing puddle. You can make your own still body of water by simply leaving a few buckets of water outside.

Collect as many wrigglers or egg rafts as possible. They can usually be found in large numbers. (See Fig. 34-1.) You'll need at least 120 for this experiment to work. Place all the wrigglers (or eggs) in one of the jars filled with water and label it "mosquitoes." (Use the same water in which you found the wrigglers or rafts.) After collecting the mosquito young, fill some extra quart-sized jars with the water for later use.

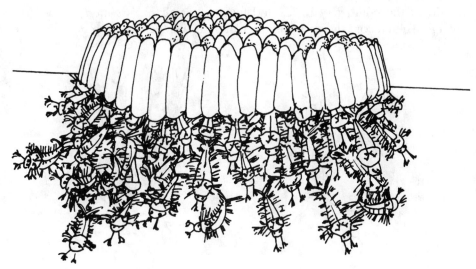

34-1 You can either use mosquito eggs or small larvae, both of which can be easily found in standing pools of water. Here young larvae are emerging from an egg raft.

Distribute the extra water so there are equal amounts of the water in each of the six jars. They need not be filled, but should have at least three inches of water. Then divide the 120 wrigglers from the jar marked "mosquitoes" into the six other jars so each has 20 mosquito young. Cut a piece of cheesecloth to act as a cover for each jar and use rubber bands to hold them tightly in place. The wrigglers cannot fly, but when they pupate and emerge as adults, they can. Keep the water in the jar marked "mosquitoes" (now empty) in case you need additional water later in the experiment.

Label three of the jars "normal" and three "warm." Place the "normal" jars in a room that maintains a temperature of about 20 degrees C (roughly 70 degrees F). Place the jars labeled "warm" either in an incubator at 35 degrees C (95 degrees F) or in a hot part of your home or school such as a boiler room. Try to keep all conditions other than the temperature the same for all jars in both groups. For example, keep them all in the dark.

Each day observe all the jars and record what you find. Continue the experiment for a few days until you notice adults flying around under the cheesecloth in some of the jars. As this happens, take the jar and place it in the refrigerator for about 5 minutes to slow the insects down. You can then use a small brush to move them into the killing jar.

Warning: *Do not inhale fumes from the killing jar.*

Leave the mosquitoes in the jar for about 10 minutes. When the mosquitoes are dead, place them into a petri or similar dish and place them under the dissecting scope or look at them with a magnifying glass. Determine the sex of each individual by first looking at their compound eyes. In most species the male's eyes come much closer together than the females as shown in Fig. 34-2. The male antennae is usually bushier than the females.

Count the number of males and females for all six jars after all the adults emerge. Then record the data and calculate the averages for each group of three.

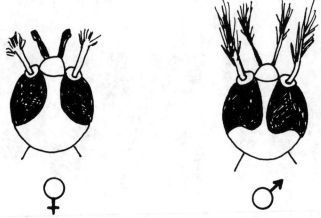

34-2 In many mosquito species, the easiest way to determine sex is to look at the eyes. Males' eyes are closer together than females.

CONCLUSIONS

Once you have the average numbers, determine the sex ratio for each group. Was there a difference in the sex ratio between the groups? Which produced the most females and males? Was the difference significant?

GOING FURTHER

If a difference was found, research why temperature changes the sex ratio.

Study the sex ratios of many different kinds of animals. Does the sex ratio for humans differ in different countries?

35
Fingernails
Growth comparisons among different groups

OVERVIEW

We all know that our fingernails grow, but do they grow faster in some people than in others? Can we make our fingernails grow faster or do they naturally grow faster during certain periods? In this experiment you will try to determine factors that affect the speed of fingernail growth. Does sex play a role? How about age or ethnic background? You can customize this experiment in any number of ways by selecting the groups you want to test.

MATERIALS

- Nail file
- Small ruler
- At least 10 people for each group to be tested (for example, if you want to compare male verse female nail growth, you need 10 males and 10 females, all about the same age)

PROCEDURES

To measure the growth of a fingernail, follow this procedure. Use the edge of a nail file to carve a fine line across the white of your thumbnail as seen in Fig. 35-1. Then, use the ruler to measure the distance from the line etched in the nail to the end of the nail, where it meets the cuticle. Each week, measure the distance from the line to the cuticle and subtract the difference from the previous measurement. This is the amount of nail growth that week. Continue taking measurements for at least one month.

Do this for each person in each group. Then calculate the average for each group to use for comparison. Fill in a table similar to Fig. 35-2.

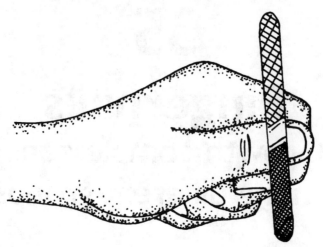

35-1 Use a file to carve a slight groove in the white of the thumb nail.

	Male group					
Name	**Nail growth in millimeters (week #)**					
	1	**2**	**3**	**4**	**5**	**Notes**
1.						
2.						
3.						
4.						
5.						
6.						
7.						
8.						

35-2 Create a table in your journal similar to this to record your observations.

CONCLUSIONS

Do your different groups have differing nail growth rates? Is so, what factors might cause these differences?

GOING FURTHER

Perform a similar experiment, but test for growth changes due to people taking vitamin supplements. For example, measure the nail growth of 10 people of similar age and health for one month while taking no vitamins. Then, have the group take a multi-vitamin for one month and repeat the nail growth test to look for differences.

Research the anatomy and physiology of nail growth.

36

It's freezing in here
Surviving freezing temperatures

OVERVIEW

Organisms that live in cold climates must either live through the winter or migrate elsewhere. Those remaining in the cold must be able to survive freezing temperatures.

Different types of animals have adapted to severe winters in different ways. Some large mammals go into a period of deep sleep (*hibernation*) where their metabolism almost stops, while some insects stop their normal development (*diapause*) until better conditions return. Many types of insects go through complete metamorphosis, meaning their bodies change forms as they mature. If one stage cannot survive through a cold winter, possibly one of the other stages can. The four stages of development include the egg, larva, pupa, and finally, adult.

Which stage (if any) of insects found near your home is able to survive freezing temperatures? Can eggs or pupae live through severe temperatures? In this experiment, you will collect eggs and pupal cases (cocoons, chrysalises, or other types of pupae), freeze them and see if they survive defrosting.

MATERIALS

- Forceps to collect the pupae
- Small model plane paintbrush (useful to collect eggs)
- Some small jars with lids to temporarily store the immature insects after being collected
- Some small jars with screw cap lids to hold the immature insects while in the refrigerator
- Marker
- Immature insect identification guide (optional)
- Cheesecloth
- Rubber bands

PROCEDURES

Have an adult go with you on a collecting field trip around your home or school in search of insect eggs (see Fig. 36-1) and pupal cases (see Fig. 36-2). Pupae include cocoons that have silky cases produced by moths, chrysalises which are bare pupae produced by butterflies, and numerous other bare pupal cases produced by insects, such as flies and beetles.

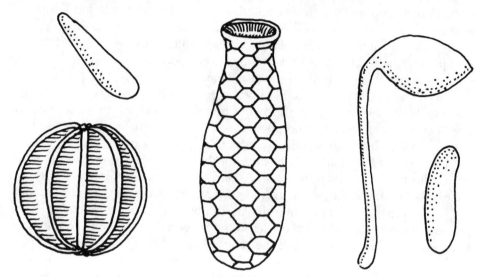

36-1 Insect eggs come in numerous varieties, but all are small and many require a magnifying glass to be distinguished.

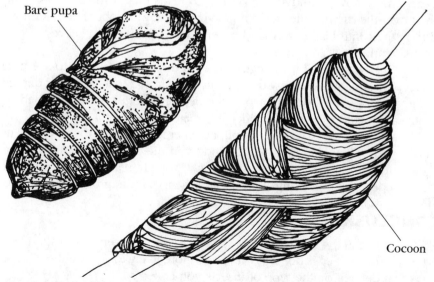

Bare pupa

Cocoon

36-2 Pupae can be enclosed in a silken cocoon or exposed with no covering.

Take along the collecting jars. Look for specimens on tree trunks, the underside of leaves, in the leaf litter, under the eaves of buildings, and in the woods. You probably already know what a cocoon looks like, but you might examine an identification guide to see what the other kinds of pupal cases and insect eggs look like.

Be very careful in removing the eggs or the pupa and placing them in the collecting jars. If they are difficult to remove, collect a small piece of the leaf, bark, or twig to which they are attached. Try to find two of each kind of pupae collected, so one can be used as the control and the other the experimental group. This shouldn't be a problem with eggs since they are usually found in groups. Be sure to take notes about where you found the specimens. This will come in handy when you try to identify them.

Once you have collected at least two types of pupae and two types of eggs (preferably more of each), return to your home or school to complete the project. (If you cannot find eggs and pupal cases, they can usually be purchased live from scientific supply houses listed in the back of this book.)

Use the identification guide to find out the kinds of eggs and cocoons you have collected. Read about these insects and see how long it normally takes for the eggs to hatch and the pupae to emerge as adults. Document all your research.

Divide the collection into two similar groups. Place one set of eggs and pupae into one jar and the matched set in the other jar. Be sure not to crush the eggs. Mark one jar "Control" and the other "Freeze." Place the "Control" group in a dark place at room temperature. (They are kept in the dark since the other jar will be in the darkness of the refrigerator.)

Place the "Freeze" group in the refrigerator for the first night and then in the freezer for the next five nights. After the five nights, remove the jar from the freezer and remove the lid so the jar and contents may warm to room temperature. Cover the jar with cheesecloth held in place by a rubber band. Remove the lid from the control jar as well and cover it with cheesecloth. Leave the opened jars in a warm room to defrost.

The only way to tell if the eggs survived the cold is to wait to see if larvae emerge. This might take a few days, many weeks, or possibly longer depending on the type of insect. (It might be considerably shorter than described in the identification guide since you artificially created a "winter.")

You can also wait to see if the pupae emerge as adults. If you prefer not to wait, however, you can cut open a window in the pupal case and touch a pin to the pupa. If the insect is still alive, it will usually react by moving. First, try this procedure with the control group and then with the group that was frozen.

CONCLUSIONS

Did the pupae in the control group survive? Did the frozen group survive? How did both groups of eggs do? Was one stage more likely to survive the winter? Answers will depend on the type of insects you collected.

GOING FURTHER

Modify the experiment to see how larvae and adults do in a freeze test.

Go on a field trip in the late fall. Find as many eggs, larvae, pupae, and adult insects as possible. You don't have to collect them, just note the kind and numbers of each found. What is the most commonly found form just before the winter hits? What does this tell you?

37
Growth spurts
The age of
fish & growth periods

OVERVIEW

Most people are familiar with the method used to tell the age of a felled tree—you count the rings on a cross-section of the trunk. You can also use these rings to indicate the amount of growth that occurred. The wider the rings, the more growth that occurred that season.

Is there a similar method you can use to estimate the age of an animal? In this project you will use fish scales to determine the age and seasonal growth in fish. Can you determine how old a fish is by looking at its scales? In addition, can you tell if the fish grew more rapidly during certain periods than others?

MATERIALS

- A few fish scales from different kinds of fish (you can get these from any fish store)
- Tweezers
- Magnifying glass or dissecting scope
- Black paper
- Tissue paper
- Envelopes

PROCEDURES

Go to the fish store and ask a sales person if you (or they) can remove a few scales from a few different types of fish. Place the scales from each type of fish into an envelope. Write the name of the type of fish on a corner of the envelope so you can identify them later.

Bring the scales back to your school or lab. Dry the scales with the tissue paper and place the scales from the first envelope on the black paper. Use the magnifying glass or dissecting scope to look carefully at the scales. Look for markings as you see in Fig. 37-1. Notice the bands across the scales. There may be two distinct types of bands, one wider than the other. Count the number of wide bands to determine the age of the fish. Research whether this technique is an accurate method of telling the age of fish. Finally, measure the thickness of the two kinds of bands. The wider the band, the more growth occurred that season.

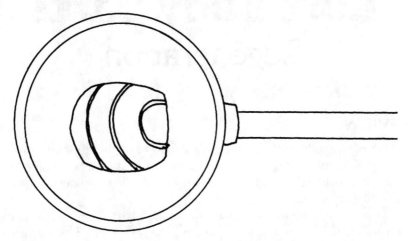

37-1 Some fish scales contain growth rings.

CONCLUSIONS

Can you tell the age of fish by counting the number of bands on the scales? Does this work for all fish? Do fish appear to grow continuously throughout their lives or in growth spurts, as indicated by bands of varying width?

GOING FURTHER

Research why the bands exist. What causes them to appear?

See if you can find any other part of an animal that indicates the age of that animal.

38
Lost body parts
Regeneration

OVERVIEW

Some invertebrate animals (without backbones) have the ability to grow back (*regenerate*) parts of the body lost in an accident or in combat. Sometimes the amputated part is large enough to form an entirely new organism, resulting in two individuals being generated from one.

Can you observe the regeneration of a single planarian resulting in two totally new individuals? How must the planarian be severed to assure two new individuals? Are starfish capable of similar regeneration?

MATERIALS

- Four or five planarian (available from a scientific supply house)
- A few microscope slides
- Scalpel
- One petri dish or syracuse dish for each planarian
- Dissecting microscope or a magnifying glass
- Supply of clean pond water or bottled spring water
- Eyedropper
- Forceps

PROCEDURES

Use the eyedropper to remove the planarian from its container and place it on a microscope slide in a drop of water. Place the slide in the refrigerator for 1 minute to slow down the planarian. While under magnification, use the scalpel to cut the planarian in half longitudinally (lengthwise) as seen in Fig. 38-1.

Use the eyedropper to place each half of the severed planarian into its own syracuse dish, or bottom of a petri dish that has been filled with fresh pond water. Repeat this procedure with another planarian. Place all the dishes containing the halved planarian into a dark room and change the water every day. Be careful not to harm the organisms while changing the water.

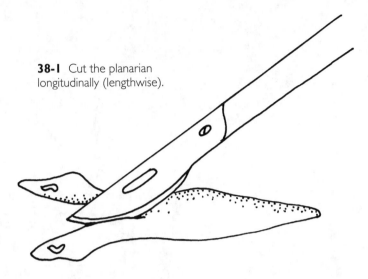

38-1 Cut the planarian longitudinally (lengthwise).

Each day, examine the planarian under a dissecting microscope and look for evidence of regeneration. Draw sketches as soon as changes appear.

CONCLUSIONS

Did the missing parts begin to regenerate in any or all of the organisms? If so, how long did it take? Did you end up with two individuals from what began as one?

GOING FURTHER

Perform a similar experiment, but cut the organisms in different directions such as transversely. Do you still end up with two individuals? Are there certain ways of cutting the animal so it does not regenerate?

Modify this experiment to see how temperature or other factors might affect the regeneration process.

Research how the regeneration you observed in the planarian differs from another invertebrate such as a starfish? (See Fig. 38-2.) Can part of a starfish regenerate into a new starfish?

38-2 The starfish is more complex than the planarian, but still capable of regenerating body parts.

Part eight

Form & function

What if you lived in a land inhabited by people with hands that look like nutcrackers, others with hands like fish nets, and still others with hands like spoons? You will learn about these imaginary people and the process of natural selection in the first project in this section.

The second project looks at the form and function of insect mouthparts. The diversity of their mouths is almost as amazing as the numbers of insects that exist. The third project investigates the physical characteristics of bioluminescence. Why is it called cold light? Does mother nature make our lightbulbs look obsolete?

In the third project, you determine which kind of natural insulation works best and in the final project you'll learn why some of us can roll our tongues and others cannot.

39

In the land of nuts, fish, & honey
Natural selection & evolution

OVERVIEW

DNA is the genetic blueprint that controls what an organism looks like. If this blueprint is changed in any way, it is called a *mutation* (a mistake). If numerous mutations occur over long periods of time and these new genes are passed on from one generation to the next, organisms won't look like their ancestors.

When these genetic mistakes occur, they sometimes make the individual more likely to survive than those without the mistake. This process is called *natural selection* or *survival of the fittest*. Since those with the genetic mistake survive, they pass this new genetic blueprint on to their offspring, so they too can take advantage of this mistake.

Let's look at an imaginary and exaggerated example of how genetic changes and natural selection might work. This imaginary world is inhabited by three species of humans who evolved different shaped hands. These three kinds of humans inhabit three different regions of the planet separated from each other by mountains.

One group of humans lives in an area where nuts are found growing everywhere. Their right hand is in the shape of a nutcracker. Another group lives in an area where fish are readily available. They have their right hand in the shape of a fish net and live primarily by catching and eating fish. The third group lives in a land where honey can be found in pools. Their hand is in the shape of a spoon.

One day an enormous sheet of ice began to cover the entire planet, including the regions where the nutcracker, fish net, and spoon hand people lived. The ice destroyed all the foods except the nuts. The fish net and spoon people began to starve since there was no longer any food they could eat.

The ice, however, formed bridges over the mountains so the fish net and spoon people crossed to the land of nuts. All three groups tried to survive in the land of nuts together.

In this project you will try to determine why there were populations with three different types of hands in the first place and what probably happened to the three groups of people at the end of the story.

MATERIALS

- A few walnuts
- One goldfish
- Small jar of honey
- Nutcracker
- Small fish net
- Few plastic spoons
- Container with water to hold the fish (a small aquarium is fine)
- Stopwatch or watch with a second hand

PROCEDURES

Have one person hold the stopwatch and time each of the following three events. First, you will play a fish net person. With the fish net in your hand attempt to 1) open a walnut, 2) catch a fish, and, 3) finally, try to eat the honey. (See Fig. 39-1.) (Throughout this experiment, do not actually eat the honey. Just see if it would be possible to do so.)

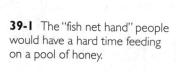

39-1 The "fish net hand" people would have a hard time feeding on a pool of honey.

Next, you will play a nutcracker person (Fig. 39-2). Hold a nutcracker in your hand. Repeat the three steps stated above. Again, time the events. (Do not harm the fish! It should be considered impossible to catch the fish with the nutcracker.) Finally, play a spoon person by repeating the three steps with a spoon. Time the events.

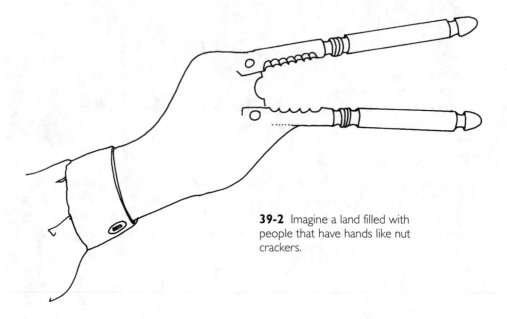

39-2 Imagine a land filled with people that have hands like nut crackers.

CONCLUSIONS

What conclusions can you draw from these experiments? Can one "tool hand" be used efficiently to obtain more than one food? If so, is it as efficient at obtaining a food that does not "naturally" correspond to it?

Think about the following questions. 1) Why do you think three different populations of hand people developed in the first place? 2) After the ice sheet covered the land and destroyed all the honey and fish, who would have found enough food to survive? 3) Would the answer to this question be the same if the fish and honey gradually disappeared instead of immediately?

GOING FURTHER

How do you think this project relates to the destruction of natural habitats that is occurring all over the world?

Read about urbanization and its effect on natural habitats.

Read more about the theory of evolution and natural selection.

40
Diverse mouths
Form & function

OVERVIEW

The mouthparts of insects are diverse and complex. Many insects not only eat with their mouthparts, but feel, smell, hold, and taste with them. Different kinds of insects have different kinds of mouths. There are sponging/lapping, chewing, piercing/sucking, and many others. Can you identify the type of mouthparts an insect will have by the food that it eats?

In this experiment, you will first become acquainted with the different types of mouthparts found in insects. Then, you will find insects and hypothesize the type of mouthparts it probably has, before catching the insect and identifying the actual mouthparts.

MATERIALS

- Insect book (entomology text) that describes and illustrates the diversity of insect mouthparts
- Dissecting microscope or a good quality magnifying glass
- Fine forceps
- Fine tweezers
- Small jars with screw caps to collect insects (for example, baby food jars)
- 70% alcohol to preserve the insects
- Marker

PROCEDURES

Study an entomology textbook to learn about insect mouthparts. Be sure to read about and see illustrations of chewing mouthparts (see Fig. 40-1), piercing/sucking, sponging/lapping mouthparts, and others. Create your own mouthpart identification guide based on the types of food an insect eats and the damage it causes to what it eats. After you understand the form of these mouthparts, have an adult take you on a field trip and look for insects.

The vast majority of insects that are not in flight will be eating. Find an insect that is munching away on a leaf, leaving large portions of the leaf missing. (See

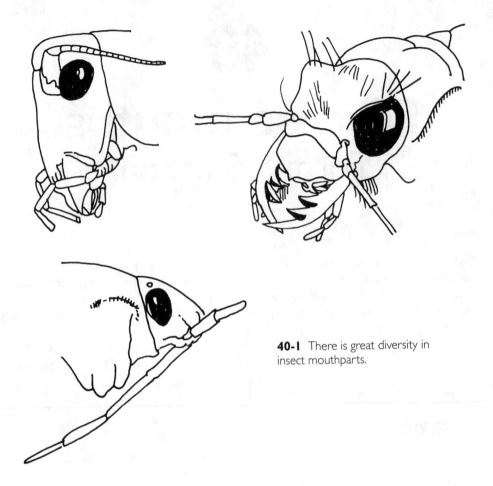

40-1 There is great diversity in insect mouthparts.

Fig. 40-2.) Collect the insects using forceps and place them in a small jar filled with 70% alcohol. Take notes indicating where they were found, what they were feeding on, and the damage caused. Mark the jar so it can be referenced to the proper notes. Write in your field book the type of mouthparts you believe it has.

Continue searching for other insects such as butterflies, beetles, different kinds of flies, leafhoppers, and others. As you find each specimen, record your notes and collect the insect. (Butterflies need not be placed in a jar. Instead, they can be temporarily stunned by a sudden squeeze on the bottom of the thorax. You can observe their mouthparts in the field and let them go.)

After collecting about a dozen different types of insects, bring them back to your school lab and look at the mouthparts under the dissecting scope or magnifying glass to see if your mouthpart identification guide works.

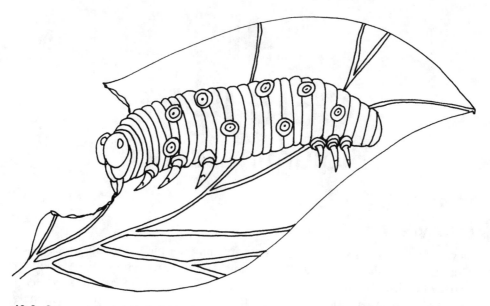

40-2 Can you tell what kind of mouthparts the insect has by looking at the damage they caused?

CONCLUSIONS

How diverse are the mouthparts of different kinds of insects? Does the form of the mouthparts follow the function of how the insect eats? Can you identify the mouthparts of an insect based on the way it was feeding without the benefit of an insect identification guide?

GOING FURTHER

Prepare permanent microscope slides of the different kinds of insect mouthparts.
    Read about unusual mouthparts such as rasping/lapping.
    Read more about form and function.

41
Cold light
Bioluminescence

OVERVIEW

As you look out across a field during a warm summer night, you might see a dazzling display of flashing lights. While commonly called fireflies, these "flies" are actually beetles. An animal's ability to produce light is called *bioluminescence*.

The firefly uses bioluminescence to find and attract a mate. The female, which cannot fly, remains on the ground flashing her light in a sequence unique to her species. Males flying nearby respond by flashing a similar sequence allowing the two to find one another.

Only a small amount of the electricity used to power a light bulb actually produces light. Most of this energy is lost as heat. This is why a light bulb is so hot if it is touched. Does the bioluminescence created by a firefly produce heat?

MATERIALS

- At least 25 fireflies, preferably more (in many areas, you can catch them on warm summer nights by simply cupping your hands around them and placing them in the jar) (See Fig. 41-1.)
- Two similar quart-size jars with screw caps
- Two similar thermometers that will fit into the jars

PROCEDURES

Place one thermometer gently into each jar. Record the temperature of both jars. (They should be the same.) Catch about 25 fireflies and place them all in one of the jars. The other jar will be the control. Place the lids tightly on the jars. Keep the jars in the dark and under similar conditions so the fireflies continue to flash.

Wait 30 minutes and then read the temperature on the thermometers again to see if the jar with insects is any warmer than the control jar.

If there was a slight increase in temperature, repeat the experiment, but place both jars (experimental with insects and control without) in light. This will reduce the flashing activity of the fireflies. Once again, after 30 minutes, read the temperature in both jars to see if there is still a difference in temperature between both jars. Release the insects outdoors when you complete the experiment.

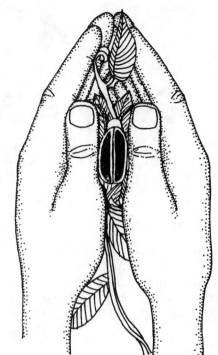

41-1 Fireflies can easily be captured by cupping your hands over the beetle.

CONCLUSIONS

Was there any difference in temperature between both jars in the first part of the experiment? If not, what does this tell you about bioluminescence in fireflies? If there was a slight difference in temperature in the first part, what happened when the flashing activity was reduced in the second part of the experiment when there was little flashing activity? Was there still a difference in temperature between the two jars? What does it mean if there was still a difference in temperature, even though the insects are no longer flashing?

GOING FURTHER

Read about the types of devices that can detect extremely small differences in temperatures. What kind of equipment would a university use to conduct a similar experiment?

Research bioluminescence and see why it is often called "cold light."

Study other animals that use bioluminescence and learn how it is used.

42
Cozy & warm
Insulation

OVERVIEW

Different kinds of animals surround their bodies with different types of insulation. Some have hair or fur, some feathers, or some blubber. In many cases, this helps the animal maintain its body temperature and reduce heat loss. In this project, you'll try to determine which type of insulation is best at retaining heat. Is human hair, down feathers, nondown feathers, or fat the best insulation?

MATERIALS

- Box of zipper baggies
- Refrigerator
- Five similar thermometers (you can do this experiment with only one thermometer)
- Handful of human hair (you can get this from a beauty parlor or barber)
- Handful of down feathers (possibly from an old down feather pillow)
- Handful of nondown feathers (possibly from an old regular feathered pillow)
- Small container of lard or shortening (you can get this from a supermarket or butcher)
- Stopwatch or watch with second hand
- One pair of disposable latex gloves
- Marker

PROCEDURES

Label the first zipper baggy "human hair." Then, fill this baggy about two-thirds with human hair. (Wear disposable gloves while handling the hair.) Place a similar amount of the other types of insulation (two types of feathers and lard) into their own labeled baggies. Label a final baggy "control," which will contain a thermometer, but no insulation.

Before placing the thermometers into the baggies, leave them out until they all reach the same room temperature. Then, insert one thermometer into the middle of each type of insulation. Place the thermometer so it is completely surrounded by the insulation and not touching the baggy.

You might want to wrap a rubber band around the baggy to hold everything in place. (See Fig. 42-1.) Be very careful placing the control thermometer into the baggy, since there is no insulation to cushion it. If you only have one thermometer, simply use one type of insulation at a time and then repeat the entire procedure for the others.

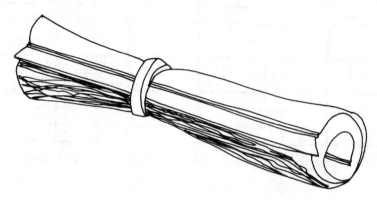

42-1 Roll the insulation-filled zipper baggy around the thermometer and hold in place with a rubber band.

With the stopwatch ready, place all the bags simultaneously into the same portion of the refrigerator about 2 inches apart, close the door, and begin the stopwatch. (Have a friend help with the next procedure.) After 10 minutes, remove the bags, and read the thermometers. Record the temperatures of each and return the thermometer to the bag and the bag to the refrigerator. Repeat this procedure after 20 minutes and after 30 minutes. Fill in a table similar to Fig. 42-2 as you proceed.

CONCLUSIONS

Once you have collected all the data, create a graph that plots the change in temperature of all insulation tested. How do they compare? Did the various forms of insulation all retain heat better than the control? Which insulated the best? How much better did it perform than the others?

GOING FURTHER

Try different types of insulation. How do manmade forms of insulation compare with feathers and fur? Are they better insulators than those created by mother nature?

Research which types of animals use which types of insulation. Research information about the natural history of the animals that use the insulation.

Type of insulation	Temperature reading				Notes
	Start	10 min	20 min	30 min	
1. Hair					
2. Down					
3. Feathers					
4. Lard					
5. Control					

42-2 Create a table in your journal similar to this to record your observations.

43
Widow's peaks & rolled tongues
Patterns of inheritance

OVERVIEW

Each of your parents contributed exactly one-half of the information that went into creating you. This information was found in the chromosomes of your father's sperm and your mother's egg. Chromosomes are rod-shaped bodies that contain genes. These genes act as a blueprint to life. The information found in these genes determines what we look like. Each gene on the chromosome affects a trait. (The word *trait* refers to some aspect of an individual, such as eye color.)

The sperm and the egg contain twenty-three chromosomes each. When the egg is fertilized by the sperm, these chromosomes combine to form twenty-three pairs. The genes on the chromosomes of the sperm have matching genes on the chromosomes of the egg. In other words, a trait is determined by a gene from the mother as well as matching gene from the father.

Each trait has many possible ways of showing itself (expressing itself). Eye color, for example, is a trait that can be expressed as black, brown, blue. Some expressions of a trait are "dominant" over other expressions. For example, brown eyes are dominant over blue eyes. Even though both of the genes in a matched set are involved in how a trait is expressed, one gene overpowers the other showing itself and hiding the other. The gene that shows itself (overpowers the other) is said to be *dominant* and the gene that is hidden is said to be *recessive*. Recessive genes only express themselves if both of the genes in the matched set are recessive, with no dominant gene hiding them.

If one were to look at large populations of people for different traits (such as eye color, hair color, attached earlobe versus unattached earlobe), which expressions are most often seen? How often will you find a recessive trait expressing itself and how often will you find a dominant trait expressing itself?

MATERIALS

- Large group of people to observe (at least 30, preferably more)

PROCEDURES

Find as many people to observe as you can (30–60 people is a good number). The larger the number of volunteers you have the more reliable your results will be. Try not to pick people all from the same family since family members tend to have similar characteristics, which would reduce the accuracy of your results.

The list below shows the traits you will be examining, the different ways these traits appear (are expressed), and whether a trait is dominant or recessive. Make a list of how many people have each of the possible expressions of each trait listed. Calculate what percentage of the total number of people observed had each particular expression of a trait.

Inherited characteristics	Genetic factor
Hairline: (see Fig. 43-1)	
Widow's peak	dominant
Continuous hairline	recessive
Earlobes:	
Unattached earlobe	dominant
Attached earlobe	recessive
Freckles:	
Freckles	dominant
No freckles	recessive
Tongue rolling: (see Fig. 43-2)	
Ability to roll tongue	dominant
Inability to roll tongue	recessive
Tongue folding:	
Ability to fold tongue	dominant
Inability to fold tongue	recessive
Eye color:	
Brown eyes	dominant
Blue eyes	recessive
Hair color:	
Dark hair (black or brown)	dominant
Light hair (blond or red)	recessive

CONCLUSIONS

After collecting all the information from your sample population, analyze the data in a table. Calculate the percentages found for each expression (blue or brown) of a trait (eye color). What conclusions can you draw from your observations? Are recessive or dominant characteristics consistently found more frequently? Are the percentages predictable?

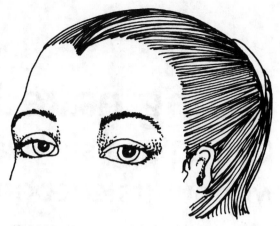

43-1 What percentage of the population has a widow's peak hairline?

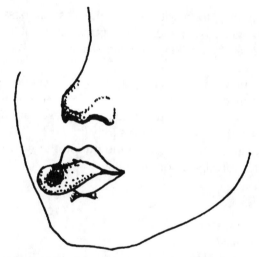

43-2 What percentage of the population can roll their tongues?

GOING FURTHER

Continue this project by researching and studying the genotypes and phenotypes of the sample population.

Study more about inheritance.

Research inherited diseases.

44
Going naked
Can a pupa survive without its cocoon?

OVERVIEW

Butterflies and moths, along with many other insects, pass through four stages as they mature: egg, larva (caterpillar), pupa, and adult. The pupal stage of moths is surrounded by a silken wrapping called a *cocoon*. This cocoon serves many functions.

Can a pupa survive in the wild with part of its cocoon damaged? What will happen to the immature insect if the cocoon becomes damaged? In this experiment, you will find cocoons in the wild and remove varying degrees of the cocoon casing to see the effect on their survival.

This project can be done at any time during the warm weather, but is best performed in the early spring or late fall.

MATERIALS

- Fine dissecting scissors
- Magnifying glass
- Immature insect identification guide
- Sharp scissors

PROCEDURES

First, look through an immature insect guide to see what most cocoons look like and where to find them. Then, have an adult take you on a field trip in search of cocoons. They can be found in the woods, on the bark of trees, in the leaf litter, and on leaves and stems. They can also be found on tall grasses, in untended weeds, along the foundations, and under the eaves of buildings.

Bring a note pad and pencil with you since you will be taking many notes to identify where you found the cocoons and what action was taken with each. Also take the scissors. Look for many of the same types of cocoon. Try to find at least six of the same kind of cocoon. As you locate each individual, be sure to record its location so it can be located at a later date.

After you've located about six similar types of cocoons select two that will be left untouched to act as the control group. Simply mark their location in your notebook stating they are the controls. Draw sketches of these individuals and take notes about their descriptions.

For the next two, use the scissors to carefully cut out of the cocoon a window that covers almost an entire side of the cocoon as you see in Fig. 44-1. (Be careful not to harm the pupa inside.) Do not remove the cocoon from where it is attached. Mark the location of these two individuals in your notebook, draw sketches, and take notes for future reference.

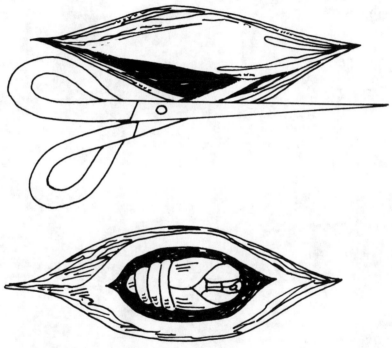

44-1 Cut a window out of the cocoon that runs its entire length.

For the remaining two, cut out a window that removes only one-quarter of a side of the cocoon. Once again, do not remove the cocoon from its place of attachment, draw sketches and take notes.

Return to each of the six sites at least once every few days. Take notes and draw sketches if changes occur. Continue your observations until obvious changes occur. If possible, continue until some of the adults emerge from the control pupal cases.

CONCLUSIONS

Did any or all of the adults emerge? What happened to the pupae that had the cocoons exposed? Did the elements, predators, or dehydration affect them? Did any survive? What can you conclude about the purpose of the cocoons?

GOING FURTHER

Research why many pupae do not have a cocoon surrounding them?

Perform a similar experiment, but apply some form of covering over the exposed window. For example, apply a plastic wrap, or aluminum foil over the opening. What happened to the pupae?

Research silk. What properties does it have and how have people used it to their advantage?

Part nine

Applied science

There's been a murder and footprints have been found near the body! "What does this have to do with zoology?" you ask. See the first project in this section to find out. The second project investigates how we have used nature's miracle drugs (antibiotics) to our benefit. In the third project, you'll see how we can use some animals to control others. This project will introduce you to the biological control of pests, a field just beginning to explode with possibilities.

The fourth project in this section tries to establish a relationship between smoking and exercise, then continues to attempt to link passive smoking (inhaling someone else's smoke) and exercise. The final project continues with the idea of biological control, but this project uses an animal to control plant pests.

45
Telltale footprints
Determining height from footprints

OVERVIEW

Police often use what might seem to be strange types of evidence to identify a criminal. One such type of evidence is the use of a person's footprints to determine height. Can you determine how tall a person is by looking at footprints? Is it possible to determine the height of our ancestors (early man) based on ancient footprints?

MATERIALS

- Yardstick
- 12" ruler
- Sandbox or area of soft earth or sand in which a footprint can be detected
- Early human (hominid) footprints—optional (you might be able to get permission from a museum or a college to use their specimens for this experiment)
- At least five friends to help you, preferably of varying heights

PROCEDURES

Smooth out the sand or soft earth over an area large enough to make a footprint. Walk through this area using your normal stride. Measure the length of the footprint from heel to toe with the ruler. (See Fig. 45-1.) Record the length. Smooth the area out and have two friends repeat this procedure making their own footprints.

Next, measure the height of the two people who just made the footprints and your own height. Then, calculate the ratio of the footprint length to the height for each person as you see in Fig. 45-2. Once you get a ratio (percentage), you will use this percentage to hypothesize the height of your three other friends.

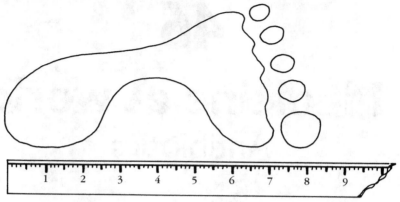

45-1 Use a ruler to measure the length of the footprint from heel to toe.

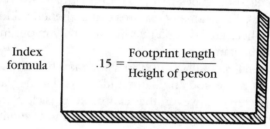

45-2 Calculate the footprint to height ratio.

Do this by having the next person walk in the sand in the same way you and your other friends did earlier. Measure the length of their footprints. Then, use their footprint length and the ratio (percentage) you determined earlier to estimate their height. Once your estimate has been made, measure the person's height to see if you were correct. Repeat this procedure for the remaining participants.

CONCLUSIONS

What did you determine the footprint length to height ratio to be for the first three people tested? Was it relatively similar for each? Once you determined the ratio, did it allow you to estimate the height of your other friends?

GOING FURTHER

Try to get permission to use preserved footprints of early humans at a museum. After receiving permission, measure the footprint length and use your ratio to determine the height of these early humans. How tall do scientists believe that these individuals were? Was your hypotheses the same as theirs?

Read about research done on early humans by Mary Leakey.

Research how footprint length and stride can help scientists (and detectives) determine sex, weight, and speed of travel.

46

Medicine at work
Antibiotics

OVERVIEW

Bacteria are one-celled microbes that are found virtually everywhere, including on and in our bodies. These small creatures can be beneficial in some cases and harmful in others. There are bacteria that live naturally in our intestinal tract and help provide us with vitamins.

There are bacteria naturally found in your mouth, throat, and throughout your gut that do not cause you any harm and are said to be part of the "normal flora." There are bacteria, however, that can make us sick. Sore throats, for example, are usually caused by a population explosion of certain types of bacteria in the mouth and throat.

Antibiotics are chemicals that are produced by some microorganisms to kill other microorganisms. By manufacturing these antibiotics ourselves, we can use them to our advantage. Antibiotics are used as a medicine against microorganisms that cause disease. Some antibiotics are effective against a wide variety of bacteria (*broad spectrum*) while others are effective against a limited number of bacteria (*narrow spectrum*).

When you are given antibiotics for a sore throat, are all the bacteria that make up the natural flora of your mouth and throat in danger of being killed? Or are only those microbes responsible for causing the sore throat killed?

MATERIALS

- A few nutrient agar plates (they can be mixed or come prepared; both are available from a scientific supply company)
- At least two different antibiotic discs and a control disc (these can be purchased from a scientific supply house)
- Forceps
- Sterile swabs
- Incubator (optional)
- Ruler

PROCEDURES

Take a sterile swab and roll it over your tongue making sure to cover the entire swab. Take this swab and rub it back and forth gently over the surface of a nutrient agar plate so you cover the entire plate. Try not to gouge the agar. Open and close the plate quickly while swabbing the agar.

Antibiotic discs are saturated with an antibiotic. The discs should come with information about the type of antibiotic they contain. Once again, open the plate and place at least two different kinds of antibiotic discs and a disc that has no antibiotic on it (this will serve as a control) on the agar so they are equidistant from one another as you see in Fig. 46-1. Mark the top of the plate with the types of antibiotic and the control. Cover the agar plate and tape it shut. Put it in an incubator or a warm room in your home. Look for growth on the agar after 24-48 hours.

Look for areas surrounding the discs that have no growth as you see in Fig. 46-1. Be especially aware of the areas around the discs that look like halos. These halos indicate no growth. Draw illustrations of the colonies.

Caution: *Do not open the lids of the agar plates.* Speak with your teacher about the proper way to dispose of these plates.

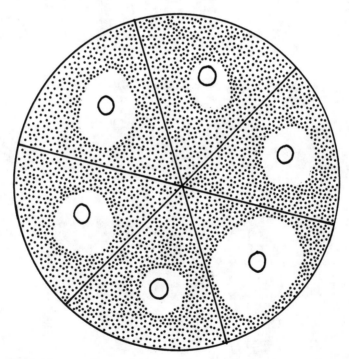

46-1 Use a marker to divide the plate into sectors with one sector for each antibiotic disc. Notice halos of no growth around some of the discs.

CONCLUSIONS

Did you find colonies growing on the agar plates? What effect did the antibiotic discs have on the growth of the bacteria around the discs? What effect did the control disc have?

Measure the halo area around the discs by placing a ruler on the lid. Were there reduced numbers of colonies around the discs or no growth at all? When you take antibiotics, what do you think is happening to the normal flora of your mouth and throat?

GOING FURTHER

To continue this experiment, look at what effect varying concentrations of antibiotics have on bacterial growth. Will a higher concentration of antibiotic kill more bacteria?

How are some antibiotics similar in some ways to pesticides?

47

Biological control
Using beneficial
insects to control pests

OVERVIEW

Farmers use tons of dangerous pesticides every year on food crops to control insect pests. Unfortunately, these pesticides not only kill pests, but many other forms of life. In nature, most animal populations are controlled by other animals. Predators attack their prey and control the population naturally.

Is it possible to select a certain predator and use it to control a certain pest? In this project, you'll see if a ladybird beetle (ladybug) can successfully control a common insect pest (aphids) on houseplants.

MATERIALS

- A few ladybird beetle larvae which can be caught in the warm summer months or ordered from a scientific supply house (See Fig. 47-1.)
- Colony of aphids (instructions are given in the Procedures section.)
- At least two, preferably four, potted bean plants all the same size
- cheesecloth

PROCEDURES

You can collect the ladybird beetle larvae using a sweep net in an open field with knee high vegetation, or order them from a supply house. (See Fig. 47-1.) You need two or three individuals for each potted plant used. (You can use the adult ladybird beetles, but they might fly away from your experiment. The larvae cannot fly.)

The easiest way to collect aphids is to simply place a potted bean plant in a very sunny spot during the summer. Within a few weeks, they should be naturally infested with aphids.

Once you have one or two similar sized bean plants infested with aphids and a few ladybird beetle larvae, you can begin the experiment. Place two or

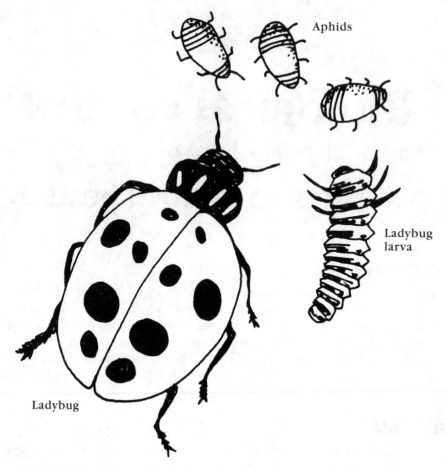

Aphids

Ladybug
larva

Ladybug

47-1 Ladybugs, which are actually beetles, prey on aphids, a common houseplant pest.

three larvae on one of the infested bean plants. Leave another, similar bean plant without the larvae as the control. Wrap the plants in cheesecloth to prevent the aphids from getting away. Keep the plants at a distance from each other. If you have additional larvae and bean plants, duplicate the experiment and the control.

Place all the pots in identical environmental conditions, water and feed the same amounts for each. Observe all the pots for activity. Do you see the larvae eating the aphids? Make observations a few times each day. Record the approximate numbers of aphids on each plant on each day. Also record the health of all the plants. Look at the color of the leaves and stems.

CONCLUSIONS

Did the larvae prey upon the aphids? If so, did they make a noticeable difference in the aphid population? Did they control the population completely? How did the health of the controls versus the experimental plants compare? Do you think ladybird beetle larvae can be used successfully to control aphids on house plants?

GOING FURTHER

Research the "biological control" of insect pests to find out how it is currently being used on a large scale.

Set up similar experiments for different kinds of predators and their pest prey.

Modify this experiment to determine how many aphids a typical ladybird beetle would devour throughout its lifetime.

Modify this experiment to see how well the larvae control the aphids compared with a common household pesticide.

48
Smoking, exercise, & the heart
Does smoking affect the pulse recovery rate?

OVERVIEW

Smoking is known to be harmful to our health. It is linked to numerous kinds of diseases, most noticeably lung cancer. Smoking affects the lives of people in many other ways as well. When people exercise, their heart beat rate increases and in turn their pulse rate increases. When a person stops exercising, the heart gradually slows back to normal speed and the pulse returns to normal. The amount of time it takes for the pulse to return to normal is called the *recovery rate*.

Is there a significant difference in the recovery rate between smokers and nonsmokers? You can continue the project to see how second hand smokers fit into the picture, as well.

MATERIALS

- Group of at least four nonsmoking adults and a similar group of four smoking adults willing to assist you in this project, all about the same age
  Caution: *Be sure all participants are in general good health and do not have heart problems.*
- Stopwatch or watch with a second hand

PROCEDURES

Have the first individual from the nonsmoking group sit and rest for three minutes. After that time, take a pulse for one minute and record the results. Then, have this person jog in place for two minutes. As soon as they finish jogging, take the pulse again for one minute and start the stopwatch. (Don't forget to record the data.)

Wait two minutes and take the person's pulse again for one minute. Record the pulse at the two minute mark. Take the pulse again every other minute, recording the pulse and the time elapsed. Do this until the pulse rate has returned to normal (the resting pulse rate).

Repeat this procedure for the other three nonsmokers and then for the smokers. Once all the data has been collected, average all the data for each group.

CONCLUSIONS

What was the recovery time for the nonsmoking and the smoking group? Was there a significant difference between the two groups? Does smoking appear to affect the ability of the heart to recover from exercise as indicated by the recovery pulse rate?

GOING FURTHER

Nonsmokers often inhale the smoke of smokers. This is called *passive, second-hand*, or *environmental tobacco smoke*. Perform the same experiment, but replace the group of smokers with a group of adults that live with a smoker.

49
Cleaning house
Snails & algae

OVERVIEW

The natural feeding habits of many organisms can be used to our advantage. Many harmful insects can be controlled by beneficial insects. In fact, bats are allowed to breed in large numbers in some cities to help control flying insect pests.

Can you use the feeding habits of a snail to control the growth of algae on an aquarium tank? How many snails would be needed to keep a tank of a certain size clean?

MATERIALS

- Four similar size tanks (they can be small, quart-sized, plastic tanks found in pet stores)
- Fresh pond water, preferably with algae growth (it will appear greenish in color)
- Seven snails of the same type and size (from a pet store)
- Marker
- Reference book about aquarium care, including information about snails

PROCEDURES

First, measure the amount of surface area on all four sides of the tanks. (Since the tanks are similar, you only need to measure one tank.) Next, identify and purchase the snails to be used and have an adult help you to collect the fresh green pond water (enough to fill the four tanks). Fill each of the tanks with the pond water to the same level. Mark the first tank "Control," the second "One Snail," the next "Two Snails," and the last "Four Snails." Place no snails in the control tank and the appropriate number in each of the others. (See Fig. 49-1.)

Place all the tanks in a warm room in indirect sunlight. All the tanks must get the same amount of light. Leave them in this position for one week. Add small amounts of additional pond water as the water evaporates.

After this period of time, observe the amount of algae growth on the sides of all the tanks. Develop a relative rating system to compare the growth. For exam-

49-1 Snails can help control algae growth.

ple, 10 means the most and 1 the least amount of growth. Record the amount of growth for each tank every few days. Continue your observations until you can determine that one (or more) of the tanks has the algae growth under control.

CONCLUSIONS

Did the control tank develop a substantial amount of algae growth? Did the other tanks appear to have the growth of algae under control? How many snails were needed to control how much surface area?

If the control does not develop significant algae growth, repeat the procedure, but place the tanks in more sunlight. If the growth in the other tanks was out of control, run the experiment again, but with more snails in each tank.

GOING FURTHER

Can you create your own guidelines for the number of snails necessary to control algae growth in various size tanks?

What other factors must be taken into consideration when using relationships between organisms to our advantage? For example, how might the snails affect the fish in the tank, or what negative qualities might the snails bring to the aquarium? Relate this to real world problems and potential solutions.

A
Using metrics

Most science fairs require that all measurements be taken using the metric system as opposed to English units. Meters and grams, which are based on powers of 10, are actually far easier to use during your experimentation than feet and pounds.

You can convert English units into metric units if need be, but it is easier to simply begin with metric units. If you are using school equipment such as flasks or cylinders, check the markings to see if any use metric units. If you are purchasing your glassware (or plastic ware), be sure to order metric markings.

Conversions from English units to metric units are given below, along with their abbreviations as used in this book. (All conversions are approximations.)

Length:
one inch (in) = 2.54 centimeters (cm)
one foot (ft) = 30 cm
one yard (yd) = .90 meter (m)
one mile (mi) = 1.6 kilometers (km)

Volume:
one teaspoon (tsp) = 5 milliliters (ml)
one tablespoon (tbsp) = 15 ml
one fluid ounce (fl oz) = 30 ml
one cup (C) = .24 liter (l)
one pint (pt) = .47 l
one quart (qt) = .95 l
one gallon (gal) = 3.80 l

Mass:
one ounce (oz) = 28.00 grams (g)
one pound (lb) = .45 kilogram (kg)

Temperature:
32 degrees Fahrenheit (F) = 0 degrees Celsius (C)
212 degrees F = 100 degrees C (Fig. A-1)

°Fahrenheit   °Celsius

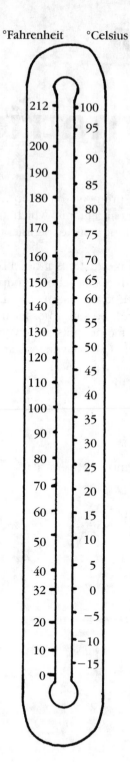

A-1 Use this thermometer to convert Celsius to Fahrenheit and vice versa.

B

Scientific
supply houses

You can order equipment, supplies, and live specimens for projects in this book from these companies.

Ward's Natural Science Establishment, Inc.
5100 West Henrietta Rd.
Rochester, NY 14692
(800) 962-2660

OR

815 Fiero Ln.
P.O. Box 5010
San Luis Obispo, CA 93403
(800) 872-7289

Blue Spruce Biological Supply Co.
221 South St.
Castle Rock, CO 80104
(800) 621-8385

The Carolina Biological Supply Co.
2700 York Rd.
Burlington, NC 27215
Eastern US: 800-334-5551
Western US: (800) 547-1733

Connecticut Valley Biological
82 Valley Rd.
P.O. Box 326
Southampton, MA 01073

Fisher Scientific
4901 W. LeMoyne St.
Chicago, IL 60651
(800) 955-1177

Frey Scientific Co.
905 Hickory Lane
P.O. Box 8101
Mansfield, OH 44901
(800) 225-FREY

Nasco
901 Janesville Ave.
P.O. Box 901
Fort Atkinson, WI 53538
(800) 558-9595

Nebraska Scientific
3823 Leavenworth St.
Omaha, NE 68105
(800) 228-7117

Powell Laboratories Division
19355 McLoughlin Blvd.
Gladstone, OR 97027
(800) 547-1733

Sargent-Welch Scientific Co.
P.O. Box 1026
Skokie, IL 60076

Southern Biological Supply Co.
P.O. Box 368
McKenzie, TN 38201
(800) 748-8735

Glossary

abstract A brief, written overview that describes your project, usually less than 250 words and often required at many fairs.

aerobic Refers to the presence of oxygen.

agar Gelatin-like substance from seaweed. Used as a solid support for microbial cultures.

algae Single-celled organisms that contain chlorophyll and photosynthesize. Some live in large colonies and are macroscopic.

amoeba A common protozoan (singled-celled animal).

anaerobic The absence of oxygen.

autoclave A machine that sterilizes objects by combining moist heat (steam) and pressure.

autotroph An organism that can produce its own food (by photosynthesis using the sun or from inorganic chemical energy). Also called a producer.

backboard The vertical, self supporting panel used in your science fair display. The board usually displays the problem, hypothesis, photos of the experimental set-up, organisms, and other important aspects of the project, as well as analyzed data in the form of charts and tables. Most fairs have size limitations for backboards.

bacteria A single-celled, microscopic organism that reproduces by fission and has no nuclear membrane.

biochemistry The study of biochemical reactions.

biocontrol The use of organisms to control pests, also called biological control.

biodiversity Refers to the vast diversity of organisms on our planet and implies the importance of all.

cell The basic unit of life. All cells are bags containing a liquid interior (the cytoplasm). The bag itself is the cell membrane.

cilia Tiny, short hair-like structures used by some one-celled organisms to move.

ciliates A type of Protozoa with cilia. The paramecium is a ciliate.

collecting aspirator A suction device used to collect insects.

colony A population of cells growing on a solid medium.

community All the populations living within a specified area make up a community.

computer modeling Using computers to analyze existing data to make projections about what will happen in the future.

consumer An organism that must consume (eat) its food as opposed to producers that make their food.

control group A test group that provides a baseline for comparison, where no experimental factors or stimulus are introduced.

culture [noun] Container of microbes with all ingredients necessary for their survival. [verb] Growth of microbes.

decomposer Organism able to break down dead organic material (such as the dead bodies of animals or dead plant leaves); fungi and many bacteria are decomposers.

dependent variable A variable that changes when the independent (also called experimental) variable changes. For example, if testing the mortality (death) rate of organisms living in soil exposed to pesticides, the mortality rate is the dependent variable and the pesticides are the independent variable.

desiccation The loss of all water.

detritus Refers to decomposing organic matter.

diapause A period of little or no activity used while waiting out unfavorable environmental conditions.

display Refers to the entire science fair exhibit of which the backboard is a part.

ecosystem Refers to the living (organisms) and nonliving (soil, water, etc.) components of a specified area such as a pond or forest and interactions that exist between all these components.

enzyme A protein that catalyzes (helps) a biochemical reaction to occur.

eukaryotic A type of organism whose cells have internal organelles and internal membranes, such as a nucleus. All nonbacterial organisms are eukaryotic, including the higher plants and animals.

exoenzyme An enzyme that is excreted outside of a cell.

exoskeleton The external supporting and protective structure of an arthropod such as an insect and lobster.

experimental group A test group that is subjected to experimental factors or stimulus for the sake of comparison with the control group. The experimental group is the one exposed to the factor being tested; for example, a plot of soil containing organisms exposed to varying amounts of pesticides. (See Experimental Variable)

experimental variable Also called the independent variable, refers to the aspect or factor to be changed for comparison. For example, the amount of pesticide the soaks into the soil.

flagellates A member of the Protozoa that use flagellum to move.

flagellum A tail-like structure used by microorganisms to move.

food chain A simple representation of "who eats what," represented by one-to-one relationships.

food web A representation of "who eats what" in an ecosystem showing multiple feeding relationships. In other words, all the food chains linked together.

fungus Primitive plants that cannot photosynthesize their own food. Most are saprophytic, meaning they feed on decaying organic matter.

genotype Refers to the genetic make-up of an organism as opposed to the characteristics of the organism, which is called the phenotype.

habitat Refers to the place where an organism lives; for example, an aquatic or terrestrial habitat.

habituation The gradual reduction of a response to an event such as a stimulus.

heterotroph Organisms that require an external source of organic chemical energy (food) to survive as opposed to autotrophic. Same as a consumer.

host The organism that supports the life of a parasite. For example, a dog is a host for a tick.

hypothesis An educated guess, formulated after thorough research, to be shown true or false through experimentation.

indigenous Refers to organisms that naturally live in an area, as opposed to foreign or exotic species that are introduced from elsewhere.

infection A growth of microorganisms within a host, causing illness in the host.

inoculum The starting material for a microbial culture.

inorganic matter Refers to substances that are not alive and did not come from decomposed organisms.

invertebrates Organisms with no backbones such as insects, starfish and lobsters.

journal Also called the project notebook, contains all notes on all aspects of a science fair project from start to finish.

leaf litter Partially decomposed leaves, twigs and other plant matter that have recently fallen to the ground, forming a ground cover.

macroscopic Large enough to see with the unaided eye.

metabolism The sum of the physical and biochemical reactions necessary for life.

metamorphosis The change in body form during development.

microbe A small organism visible only with a microscope. Could be a bacteria, algae, fungi, protist, or virus.

morphology The study of the appearance of an organism, including its shape, texture, and color.

mycoses Infectious diseases caused by fungi.

nematodes Also called roundworms; small unsegmented microscopic worms found in most habitats in great numbers. Most are harmless, but a few are parasitic.

nucleus A membrane-enclosed structure that contains genetic material in a eukaryotic cell.

obligate pathogens Microorganisms that must have a host to survive and reproduce. In comparison to facultative pathogens, which are accidental contaminants of a host and can survive outside of a host.

observations A form of qualitative data collection.

organelle A membrane-enclosed structure within a cell in eukaryotic organisms.

organic Refers to substances that compose living, or dead, decaying organisms, and their waste products. Carbon is the primary element.

ovipositor The external female reproductive organ of an insect used to lay eggs.

paedeogensis The ability of an immature stage of an organism to produce young.

parasite An organism that lives in or on one or more organisms (hosts) for a portion of its life. The host is usually not killed in the process.

parasitoid An animal that lives in another organism (host) and kills the host during its development.

parthenogenetic reproduction The ability to reproduce without a mate (reproduction without the fertilization of the egg).

pathogens Organisms that cause disease in other organisms.

phenotype Refers to the physical characteristics of an organism, as opposed to the genetic make-up, which is called the genotype.

populations All the members of the same species living in a specific area; for example, the population of silver foxes living in Maine.

population dynamics The study of populations and factors that affect them.

pheromone A chemical that communicates information between members of the same species.

predator An animal (consumer) that eats other animals for nourishment.

producer An organism that makes its own chemical energy (food), usually using energy from the sun.

Protista A kingdom of living things, composed of single-celled eukaryotes that do not have a cell wall. Some have chlorophyll while others do not. (A file cabinet in the example given in the Introduction of this book.)

Protozoa A group of protists that do not contain chlorophyll. Complex, single-celled animals (eukaryotes).

qualitative studies Experimentation where data collection involves observations, but no numerical results.

quantitative studies Experimentation where data collection involves measurements and numerical results.

raw data Any data collected during the course of an experiment that has not been manipulated in any way. As opposed to smooth data.

research Also called a literature search, refers to locating and studying as much of the existing information about a subject as possible.

resolving power, microscope The smallest distance between two objects in which the two objects can still be distinguished from one another. If the two objects are beyond the resolving power of a microscope, the two objects appear as one.

scavenger An organism that consumes dead organic matter.

scientific method The basic methodology of all scientific experimentation including: 1) the statement of a problem to be solved or question to be answered to further science, 2) the formulation of a hypothesis, and 3) performing experimentation to determine if the hypothesis is true or false, including data collection, analysis, and arriving at a conclusion.

species Organisms with the potential to breed and produce viable offspring. (A sheet of paper, in the example given in the Introduction section.)

statistics Refers to analyzing numerical data.

sterile The absence of life.

stimulus An "event" that prompts a reaction or a response.

survey collection A collection of organisms from a certain habitat or area.

sweep net An insect collecting net designed to be swept through vegetation to collect large numbers of insects quickly.

transfer aspirator A device that allows for the easy transfer of small insects from one area or container to another using suction.

variables A factor that is changed to test the hypothesis.

vertebrates Refers to animals with backbones such as reptiles, amphibians, birds, and mammals.

virus A package of genetic material surrounded by a protein capsule that requires a living host to reproduce.

Resources

Following are lists of books about zoology and science fair projects.
These field guides can help you identify organisms mentioned in this book:

Arnett, R. & R. Jacques, 1981. *Simon & Schuster's Guide to Insects.* New York: Simon & Schuster.

Bland, R.G. & H.E. Jaques, 1978. *How to Know the Insects.* Dubuque, Iowa: Wm. C. Brown Co. Publishers.

Booth, Ernest S., 1970. *How to Know the Mammals.* Dubuque, Iowa: Wm. C. Brown Co. Publishers.

Borror, D.J. & D.M. DeLong, 1970. *A Field Guide to the Insects. Peterson Field Guide.* Boston: Houghton Mifflin.

Borror, D.J. & R.E. White, 1970. *A Field Guide to the Insects of America North of Mexico.* Boston: Houghton Mifflin.

Burch, 1962. *How to Know Eastern Land Snails.* Dubuque, Iowa: Wm. C. Brown Co. Publishers.

Chu, H.F., 1949. *How to Know the Immature Insects.* Dubuque, Iowa: Wm. C. Brown Co. Publishers.

Dashefsky, H.S. & J.G. Stoffolano, 1977. *A Tutorial Guide to the Insect Orders.* Minneapolis: Burgess Publishing Company.

Jahn, T.L., E.C. Bovee & F.F. Jahn, 1979. *How to Know the Protozoa.* 2nd ed. Dubuque, Iowa: Wm. C. Brown Co. Publishers.

Jaques, H.E., 1946. *How to Know Living Things.* Dubuque, Iowa: Wm. C. Brown Co. Publishers.

Kaston, 1952. *How to Know Spiders.* Dubuque, Iowa: Wm. C. Brown Co. Publishers.

Lehmkuhl, D.M., 1979. *How to Know the Aquatic Insects.* Dubuque, Iowa : Wm. C. Brown Co. Publishers.

McCafferty, W.P., 1982. *Aquatic Entomology.* New York: Jones & Bartlett.

Prescott, G.W., 1978. *How to Know the Freshwater Algae.* Dubuque, Iowa: Wm. C. Brown Co. Publishers.

Schultz, 1969. *How to Know Marine Isopod Crustaceans.* Dubuque, Iowa: Wm. C. Brown Co. Publishers.

If you are new to science fairs, these books cover all aspects of entering a science fair:

Bombaugh, Ruth. 1990. *Science Fair Success*. Hillside, NJ: Enslow Publishers.
Irtz, Maxine. 1987. *Science Fair—Developing a Successful and Fun Project*. Blue Ridge Summit, PA: TAB Books.
Tocci, Salvatore. 1986. *How To Do A Science Fair Project*. New York: Franklin Watts.

The following books can all be used for additional science fair project ideas. Although not specifically about zoology, many involve animals or can be adapted to create zoological projects:

Barr, George, 1964. *Science Projects for Young People*. New York: Dover Publications.
Berman, William, 1986. *Exploring With Probe and Scalpel—How to Dissect—Special Projects for Advanced Studies*. New York: Prentice Hall Press.
Bochinski, Julianne, 1991. *The Complete Handbook of Science Fair Projects*. New York: Wiley & Sons, Inc.
Bonnet, Robert and G. Daniel Keen, 1990. *Environmental Science: 49 Science Fair Projects*. Blue Ridge Summit, PA: TAB/McGraw-Hill, Inc.
Bybee, Dr. Rodger, 1987. *Acid Rain: Science Projects*. St. Paul, MN: The Acid Rain Foundation, Inc.
Dashefsky, H. Steven, 1993. *Insect Biology: 49 Science Fair Projects*. Blue Ridge Summit, PA: TAB/McGraw-Hill, Inc.
Durant, Peggy Raife, 1991. *Prize Winning Science Fair Projects*. New York: Scholastic, Inc.
Gardner, Robert, 1989. *More Ideas for Science Fair Projects*. New York: Franklin Watts.
Harlow, Rosie and Gareth Morgan, 1991. *175 Amazing Nature Experiments*. New York: Random House.
Imes, Rick, 1992. *The Practical Entomologist*. New York: Simon & Schuster, Inc. (for the hobbyist)
Irtz, Maxine, 1991. *Blue-Ribbon Science Fair Projects*. Blue Ridge Summit, PA: TAB Books.
Kneidel, Sally Stenhouse, 1993. *Creepy Crawlies and the Scientific Method*. Golden, CO: Fulcrum Publishing.
Sheehan, Kathryn and Mary Waidner, Ph.D., 1991. *Earth Child: Games, Stories, Activities, Experiments & Ideas About Living Lightly on Planet Earth*. Tulsa: Council Oak Books.
Sisson, Edith A., 1982. *Nature with Children of All Ages*. New York: Prentice-Hall Press.
Tant, Carl, 1992. *Science Fair Spelled W-I-N*. Angleton, TX: Biotech Publishing.
VanCleave, Janice, 1990. *Biology for Every Kid: 101 Easy Experiments that Really Work*. New York: John Wiley & Sons, Inc.

VanCleave, Janice, 1993. *A+ Projects in Biology.* New York: John Wiley & Sons, Inc.

Witherspoon, James D., 1993. *From Field to Lab, 200 Life Science Experiments for the Amateur Biologist.* Blue Ridge Summit, PA: TAB Books.

For information about the International Science and Engineering Fairs and valuable information about Adult Sponsorship, write or call:

The Science Service
1719 N St., N.W.
Washington, DC 20036
(202) 785-2255

Index

Boldface numbers indicate illustrations

diurnal activity in animals, 4
diversity of animal life on earth, xvii
dominant characteristics, 135

E

earthworms
  eating preferences, 53-55, **53**, **54**
  rising groundwater levels, 77-79, **78**
eggs of various insects, **115**
environments for animals, 35-55
  earthworm eating preferences, 53-55, **53**, **54**
  food chains, feeding relationships, 43-48, **44**, **45**
  optimum levels: animal's preferred habitats, 85-87, **86**, **87**
  predators and prey animals, 40-42, **41**
  temperature vs. mosquito development, 36-39, **37**, **38**
  tolerance ranges of plankton, 49-52, **50**, **51**
evolution, 124-126, **125**, **126**
excretion of wastes from animal bodies, xviii
experimentation, xv

F

Fahrenheit to Centigrade conversion, **156**
families of animals, xix
fecundity, 80
feeding habits of animals, xvii
fingernail growth rates, 111-113, **112**
fireflies
  bioluminescence, 130-131, **131**
  communication, 90-91, **90**
fish
  circulatory system, 25-27, **26**
  growth periods, 118-119, **119**
  respiratory rate of fish vs. temperature, 82-84, **83**
food chains, feeding relationships, xvii, 43-48, **44**, **45**, **46**, **47**
footprints and height, 142-143, **143**
form and function, 123-140
  bioluminescence, 130-131, **131**
  inheritance of characteristics, 135-137, **137**
  insulation and body temperatures, 132-134, **133**, **134**
  mouth variations in insects, 127-129, **128**, **129**
  natural selection and evolution, 124-126, **125**, **126**
  pupae and cocoons, 138-140, **139**
freezing temperatures and survival, 114-117, **115**

G

genera of animals, xix
genes, 135

genetics, 135-137, **137**
geotaxis and insect behavior, 13-15, **14**
gestation, 80-81, **81**
glucose, 61-63, **62**, **63**
Going Further section of projects, xiv
graphs (*see* tables, charts, graphs)
growth and development, 103-121
  fingernail growth rates, 111-113, **112**
  fish and growth periods, 118-119, **119**
  freezing temperatures and survival, 114-117, **115**
  regeneration of lost parts, 120-121, **121**
  sex ratios, 103
    temperature vs., 108-110, **109**, **110**
  vitamins vs. cooking times, 104-107, **105**, **106**

H

habitats of animals, xvii
  optimum levels: animal's preferred habitats, 85-87, **86**, **87**
heartbeat, smoking, exercise, heart, 150-151
height from footprints, 142-143, **143**
herbivores, 43
hypotheses, xv

I

infections, 58-60, **59**, **60**
  antibiotics, 144-146, **145**
  antigens and antibodies, 64-66, **65**
ingestion and digestion of food in animals, xviii
  one-celled animals, 67-69, **68**, **69**
inheritance of characteristics, 135-137, **137**
insecticides, biological controls for insects, 147-149, **148**
insects
  arthropods, 40
  attractants for night-flying insects, 97-99, **98**
  crustaceans, 40
  eggs of various insects, **115**
  geotaxis, 13-15, **14**
  mouth variations in insects, 127-129, **128**, **129**
  pheromones and insect attraction, 2-3, **3**
  predator and prey, 40-42, **41**
  pupae and cocoons, 138-140, **139**
  species of insects, numbers, xvii
insulation and body temperatures, 132-134, **133**, **134**
iodine test for starch in food, 23-24, **24**

J

Japanese beetles, 7-9, **7**

About the author

Steven Dashefsky is an adjunct professor of environmental science at Marymount College in Tarrytown, New York. He is the founder of the Center for Environmental Literacy, which was created to educate the public about environmental topics. He holds a B.S. in biology and an M.S. in entomology, and is the author of more than ten books that simplify science and technology.